Precambrian Evolution of the North China Craton

Precambrian Evolution of the North China Craton

Guochun Zhao

ELSEVIER

AMSTERDAM • BOSTON • HEIDELBERG • LONDON • NEW YORK • OXFORD
PARIS • SAN DIEGO • SAN FRANCISCO • SINGAPORE • SYDNEY • TOKYO

Elsevier

The Boulevard, Langford Lane, Kidlington, Oxford, OX5 1GB, UK

Radarweg 29, PO Box 211, 1000 AE Amsterdam, The Netherlands

First published 2014

British Library Cataloguing-in-Publication Data

A catalogue record for this book is available from the British Library

Library of Congress Cataloging-in-Publication Data

A catalog record for this book is available from the Library of Congress

ISBN: 978-0-12-407227-5

For information on all Elsevier publications
visit our website at **store.elsevier.com**

This book has been manufactured using Print On Demand technology. Each copy is produced to order and is limited to black ink. The online version of this book will show color figures where appropriate.

Printed and bound by CPI Group (UK) Ltd, Croydon, CR0 4YY

Transferred to digital print 2012

CONTENTS

PREFACE

The North China Craton has long been a training ground for Chinese geologists, particularly structural, metamorphic, and Precambrian geologists, and a testbed for new methods and models of investigations of the Precambrian crust. Yet, despite its popularity with Chinese geologists and its world-wide reputation as one of the oldest cratonic blocks containing rocks as old as 3.85 Ga, few books written in English present an extensive and comprehensive introduction to the Precambrian geology of the North China Craton. This is partly because unlike other cratonic blocks that were intensively investigated in the 1980s and 1990s, the North China Craton was among the most poorly understood areas in the world until the beginning of this century. Since 2000, Chinese researchers and their international collaborators from Australia, Germany, United States, England, France, Canada, and many other countries have carried out extensive field-based structural, metamorphic, geochemical, geochronological, and geophysical investigations on the North China Craton, and produced a vast amount of new data and competing interpretations, which have led to major advancements in understanding the accretion and amalgamation of the craton. The most prominent achievement of these investigations is recognition of a number of micro-continental blocks and associated Himalayan-type orogenic belts in the craton. There is now a broad consensus that the North China Craton was formed mainly through Neoarchean accretion and Paleoproterozoic amalgamation of several small blocks along a few Paleoproterozoic continent—continent collisional belts, though the eastern part of the craton preserves minor Paleoarchean to Mesoarchean records. These achievements have been reflected in over 200 papers published in international peer-review journals in the past 15 years. However, most of these publications focused on single geological questions or some special issues on the formation and evolution of individual terranes or complexes in the North China Craton, and few attempted to make a synthetic overview on these new data and interpretations. This forms a justification for this book entitled "Precambrian Evolution of the North China Craton" in which I made a compilation, summary, and assessment of new data obtained for various tectonic domains in the

North China Craton, and evaluated current models proposed in the last few years. I believe such a book will not only make a timely addition to the literature on the region, but will also open a new window through which the international geological community would know what major advancements have been made in understanding the Precambrian geology of the North China Craton and what issues are still unresolved and controversial at present.

This book consists of five chapters, of which the first chapter outlines the tectonic subdivision of the North China Craton and different models proposed for the formation and evolution of the craton. Although the current tectonic models are largely mutually exclusive in terms of timing and tectonic processes of accretionary and collisional events that led to the final assembly of the craton, most models accept the subdivision of the North China Craton into the Eastern and Western blocks which amalgamated along a central collisional belt (given different names like the Trans-North China Orogen, Jinyu Belt, or Central Orogenic Belt) to form a coherent block. In addition, most models concede the existence of a Paleoproterozoic orogenic belt within the Western Block, named the Khondalite Belt, the Fengzhen Belt, or the Inner Mongolia Suture Zone, which divides the block into the Yinshan Block in the north and the Ordos Block in the south. These agreements on tectonic subdivisions of the North China Craton form a valuable basis for subsequent specialized discussion in Chapters 2–5, of which Chapters 2 and 3 focus on Archean accretion and reworking of the Eastern and Western blocks, respectively, Chapter 4 discusses the Paleoproterozoic amalgamation of the North China Craton, and the last chapter is concentrated on the Mesoproterozoic accretion and Meso-Neoproterozoic extension and rifting occurring in the North China Craton, which are considered to have been related to the outgrowth, fragmentation, and breakup of the Paleo-Mesoproterozoic Columbia (Nuna) supercontinent.

In summary, I believe that this book has highlighted the most recent research outcomes of Chinese geoscientists and a number of overseas workers who have long worked on Precambrian geology of China. I hope it will not only provide an important source of information for the international geological community but will also stimulate further research.

ACKNOWLEDGMENTS

My research on the Precambrian geology of the North China Craton has spanned a period of over 25 years, during which time many friends, colleagues, and research students have helped me in a number of different ways. I am particularly indebted to Prof. Simon A. Wilde at Curtin University and Prof. Peter A. Cawood now at University of St Andrews, who, as the supervisors of my PhD study at Curtin University, contributed immense supervision, advice, and training during the course of my PhD work by their boundless enthusiasm. Without their academic enlightenment, technical advice, financial support, and English-language help, I would not have successfully completed my PhD research at Curtin University with producing 13 peer-review papers during a 4-year PhD study. In this sense, this book should be regarded as an extension of my PhD thesis entitled "Assembly of the North China Craton: Constraints from the Tectonothermal Evolution of Metamorphic Complexes in the Trans-North China Orogen," which was completed at Curtin University in August 2000. I owe much to my best friend and colleague Min Sun who brought me to the University of Hong Kong where I initially worked as a postdoc fellow and now as a full-time professor and received a large number of research grants from Hong Kong RGC to financially support my research on the North China Craton. I also owe much to Professors Liangzhao Lu, Mingguo Zhai, Jahn Bor-ming, Qihan Shen, Guowei Zhang, Robert T. Pidgeon, Alfred Kröner, Kent C. Condie, M. Santosh, Kaiyi Wang, Songnian Lu, Timonthy M. Kusky, and many others for numerous discussions and arguments on Precambrian geology of the North China Craton. Many ideas in this thesis were initiated and rectified during these thoughtful discussions and arguments. I am particularly grateful to my long-term collaborators Jinghui Guo, Sanzhong Li, Shuwen Liu, Fulai Liu, Fuyuan Wu, Wei Jin, Xuping Li, and Chunming Wu, who have done fieldwork with me in the North China Craton for many times and inspired and energized me each time when I met difficulties; these experiences will always remain with me. I have been fortunate to have had a particularly lively group of research students, including Jian Zhang (Jason), Yanhong, He, Wing Hang Leung (Allen), Changqing Yin, Kan Kuen Wu (Philip), Chaohui Liu,

Pui Yuk Tam (Tammy), and Meiling Wu, who have worked with me on the North China Craton and produced an abundant amount of data and new interpretations that are helpful in refining and modifying my ideas. Thanks must go finally to my wife Shuke Wu and my son Geoff Zhao for their love, understanding, and moral support.

September 30, 2013

CHAPTER 1

Tectonic Subdivision of the North China Craton: An Outline

1.1 INTRODUCTION

As a general term used to refer to the Chinese part of the Sino−Korean Craton, the North China Craton (NCC) is one of a few oldest cratonic blocks in the world that contain Eoarchean rocks as old as 3.8 Ga. It covers most of North China, the southern part of Northeast China, Inner Mongolia, Bohai Bay, and northern part of the Yellow Sea, with an area of approximately 1,500,000 km^2 (Figure 1.1). The craton is bounded by Phanerozoic orogenic belts, of which the early Paleozoic Qilianshan orogen and the late Paleozoic Tianshan−Inner Mongolia−Daxinganling orogen of the Central Asian Orogenic Belt bound the craton to the west and the north, respectively, and in the south the Mesozoic Qinling−Dabie−Sulu high/ultrahigh-pressure metamorphic belt separates the craton from the Yangtze Block of the South China Craton (Figure 1.1).

The geological attraction of the NCC resides in excellent exposures of diverse basement rocks, abundant mineral resources (e.g., gold, iron, base metals, sillimanite, and graphite), complicated structural signatures, and metamorphic continuum from subgreenschist to granulite facies. Whereas much is now known of the geology of the craton, including the economic mineral occurrences, its tectonic division and evolution have not been well constrained until recently. Traditionally, the NCC was considered to be composed of a relatively uniform Archean to Paleoproterozoic metamorphic basement, partially overlain by Mesoproterozoic to Cenozoic cover, and its tectonic history was explained using a preplate tectonic geosynclinal model (Huang, 1977; Ren, 1980), which proposed that the basement rocks of the craton formed during four distinct geosynclinal cycles: Qianxi (>3.0 Ga), Fuping (3.0−2.5 Ga), Wutai (2.5−2.4 Ga), and Lüliang (2.4−1.8 Ga) (Ma and Wu, 1981). Correspondingly, four tectonothermal events, named the Qianxi, Fuping, Wutai, and Lüliang movements, were proposed to occur at ∼3.0, ∼2.5, ∼2.4, and ∼1.8 Ga, respectively

Precambrian Evolution of the North China Craton. DOI: http://dx.doi.org/10.1016/B978-0-12-407227-5.00001-8

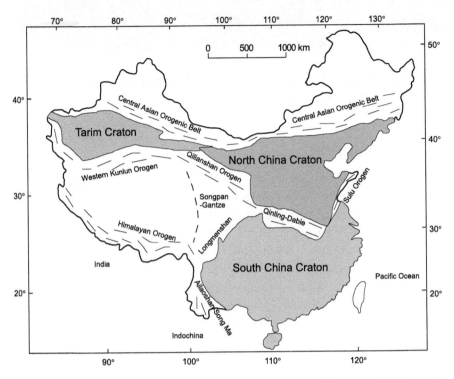

Figure 1.1 Schematic tectonic map of China showing the major Precambrian blocks connected by Phanerozoic fold belts (Zhao et al., 2001a).

(Huang, 1977; Ren, 1980; Ma and Wu, 1981). The Qianxi cycle was considered to represent the formation of some microcontinental nuclei surrounded by small-scale geosynclines in which sedimentary and volcanic rocks developed and were subsequently deformed and metamorphosed in granulite facies during the Qianxi Movement at ∼3.0 Ga (Ma and Wu, 1981). The Fuping cycle was thought to mark the development of large-scale geosynclines surrounding the earlier microcontinental nuclei (Ma and Wu, 1981). In these geosynclines, large volumes of volcanics and sedimentary rocks formed and were folded and metamorphosed from amphibolite to granulite facies during the Fuping Movement at ∼2.5 Ga, resulting in the formation of the present-day sized NCC, whereas the Wutai and Lüliang cycles were considered to represent the development of local, small-scale, intracontinental geosynclines where sedimentary and volcanic rocks formed and were deformed and metamorphosed in greenschist and subgreenschist facies, respectively, during the Wutai (∼2.4 Ga) and Lüliang (∼1.8 Ga) movements, which resulted in the further stabilization of the NCC (Huang, 1977; Ren, 1980; Ma and Wu, 1981).

The above geosynclinal model was based upon a few unconformities, K−Ar, Rb−Sr, and conventional multigrain U−Pb zircon geochronology, and assumptions that (i) the whole basement of the craton is dominated by metamorphosed sedimentary and volcanic rocks, and (ii) high-grade metamorphic rocks are always older in age than low-grade ones. However, geological investigations and mapping carried out in the late 1980s and the early 1990s have revealed that large amounts of felsic gneisses cropping out in the NCC are metamorphosed tonalitic−trondhjemitic−granodioritic (TTG) plutons (Zhai et al., 1985; Li et al., 1987; Wang et al., 1991a; He et al., 1992), not supracrustals as previously considered (Huang, 1977; Ren, 1980). Also, some so-called "unconformities" between these tectonic cycles are regional-scale ductile shear zones (Li and Qian, 1991). Moreover, new geochronological data indicate that not all low-grade metamorphic rocks are younger than the high-grade rocks, and some low-grade rocks are even older than high-grade rocks. For example, the granitoids from the low-grade Wutai granite-greenstone belt are older than the high-grade Hengshan and Fuping TTG gneisses (Wilde et al., 1997, 1998). Because of these reasons, the polycyclic geosynclinal model has been abandoned in the modern study of Precambrian geology of the NCC.

Since the 1990s, terrane accretion and collision models under the regime of plate tectonics have been applied to the NCC following the discoveries of high-pressure mafic and pelitic granulites, retrograded eclogites, fragments of ancient oceanic crust (ophiolites), mélanges, crustal-scale ductile shear zones, and large-scale thrusting and folding slices including sheath folds in the central zone of the craton (Li et al., 1990; Li and Qian, 1991; Tian, 1991; Wang et al., 1991a, 1994, 1996, 1997; Bai et al., 1992; Zhai et al., 1992, 1995; Guo et al., 1993; Zhang et al., 1994; Ma and Wang, 1995; Li and Liu, 1996; Guo and Shi, 1996). For example, based on the presence of fragments of ancient oceanic crust (ophiolites) and mélanges in the Wutai Complex, Li et al. (1990), Bai et al. (1992), and Wang et al. (1996) proposed continent−arc−continent collisional models for the evolution of the Fuping−Wutai−Hengshan area in which the high-grade Fuping and Hengshan gneiss complexes were considered as two exotic Archean continental blocks, whereas the low-grade Wutai Complex is interpreted as ancient oceanic crust and intervening island arcs which were deformed and metamorphosed during the collision between the two blocks. Alternatively, Tian (1991) argued that the Fuping and Hengshan complexes represent a single continental basement that underwent Neoarchean rifting associated with formation of the Wutai Complex and closed upon itself in the early Paleoproterozoic. Zhai et al. (1992, 1995) initially suggested that the high-pressure granulites and retrograded eclogites in the area between the Hengshan and Huai'an complexes resulted from the collision between the Huai'an and Hengshan continental blocks, though later Zhai et al. (2005, 2010) abandoned their early continent–continent model (also see Zhai, 2004, 2011; Zhai and Peng, 2007; Zhai and Santosh, 2011). Similarly, Wang et al. (1994) proposed that the high-pressure granulite belt in the Xuanhua Complex formed at ~2.5 Ga during the collision between a southern block and a northern block in the central zone of the NCC. Li and Liu (1996) argued that the crustal-scale Longquanguan ductile shear zone represents a continental suture along which the Hengshan–Wutai terrane and the Fuping terrane collided at ~2.5 Ga. Many other similar models have also been proposed for evolution of the NCC. However, most of these models were established upon local geological data from two or three complexes in the craton, and no overall synthesis of geological data from across the whole craton was undertaken. Thus, these models have not reasonably explained the tectonic evolution of the whole NCC.

1.2 MODELS FOR TECTONIC SUBDIVISION AND EVOLUTION

Since the late 1990s and early 2000s, Chinese researchers and their international collaborators from Australia, Germany, United States, England, France, Canada, Japan, etc. have carried out extensive field-based structural, metamorphic, geochemical, geochronological, and geophysical investigations on the NCC, which have led to major advancements in understanding the Neoarchean to Paleoproterozoic accretion and amalgamation of the craton. The most prominent achievement of these investigations is recognition of a number of microcontinental blocks and associated collision-type orogenic belts in the NCC. Now there is a broad consensus that the NCC was formed through the accretion and amalgamation of a number of microcontinental blocks along the collision-type orogenic belts. However, controversy has surrounded the issues of how many microcontinental blocks originally existed, and when and how these microcontinental blocks were accreted and finally assembled to form the coherent basement of the NCC. Such controversial issues have been highlighted in a number of tectonic subdivision models proposed by different research groups (Wu and Zhong, 1998; Zhao et al., 1998, 2001a, 2005; Zhai et al., 2000, 2003, 2005; Zhai and Liu, 2001, 2003; Wilde et al., 2002; Kusky et al., 2001, 2007; Kusky and Li, 2003; Kröner et al., 2005a, 2005b, 2006; Faure et al., 2007; Kusky and Santosh, 2009; Santosh, 2010; Santosh and Kusky, 2010; Kusky, 2011a,b; Zhai and Santosh, 2011; Santosh et al., 2012; Zhao and Guo, 2012; Zhao and Cawood, 2012).

Wu et al. (1998) divided the NCC into the Mengshan, Yuwan, Jinji, Qianhuai, and Jiaoliao blocks (Figure 1.2A), of which the latter two blocks were considered to have been fused to form a larger block at ~2.5 Ga, which then collided with the Mengshan, Yuwan, and Jinji blocks to form the coherent basement of the craton during the Lüliang Orogeny at ~1.8 Ga. Zhai et al. (2000) subdivided the NCC into six small blocks, named the Jiaoliao, Fuping, Xuchang, Qinhuai, Alashan, and Jining blocks (Figure 1.2B), and proposed that these six blocks were assembled to form the NCC at ~2.5 Ga. Zhang et al. (1998) even divided the basement of the NCC into 15 blocks/terranes, but they did not explain where and when these blocks were united to form a coherent craton. These models overrate lithological, structural, metamorphic, and geochronological differences between the proposed terranes/blocks, but do not well define the intervening collisional belts that resulted from the amalgamation of these blocks. Most recently, Zhai and Santosh (2011) made significant modifications on the Zhai et al. (2000) model by suggesting that the microcontinental blocks in the NCC, including the Jiaoliao (JL), Qianhuai (QH), Ordos (OR), Jining (JN), Xuchang (XCH), Xuhuai (XH), and Alashan (ALS) blocks (Figure 1.3), were welded by ~2.7 and 2.5 Ga Neoarchean greenstone belts, which were interpreted as the vestiges of arc-continent collision. Like those in other cratonic blocks, however, the Neoarchean greenstone belts in the NCC lack typical features of collision tectonics.

A number of tectonic models for the formation and evolution of the NCC have stressed the significance of continent–continent collisional

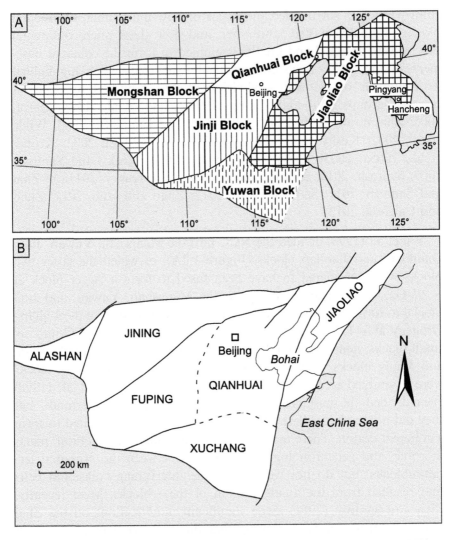

Figure 1.2 Tectonic subdivision of the NCC proposed (A) by Wu et al. (1998) and (B) by Zhai et al. (2000).

belts in identification of different continental blocks (Zhao et al., 2001a, 2005, 2007; Kusky and Li, 2003; Zhai, 2004; Kusky et al., 2007; Kusky, 2011a; Faure et al., 2007; Trap et al., 2007, 2008, 2009a, b, 2011, 2012; Zhai and Peng, 2007; Santosh et al., 2010; Zhai and Santosh, 2011). As pointed out by Zhao (2009), recognition of a continent–continent collisional belt is a key to identifying two different continental blocks. In other words, as long as two terranes are separated by a continent–continent collisional belt, they should be

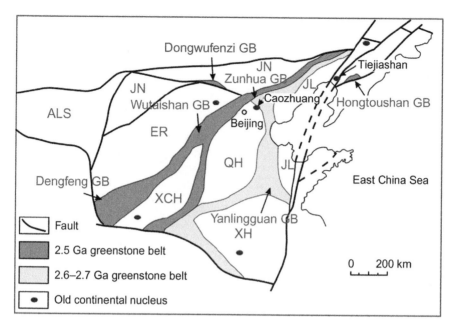

Figure 1.3 Subdivision of the NCC proposed by Zhai and Santosh (2011).

regarded as two discrete continental blocks no matter whether their lithological, structural, metamorphic, and geochronological features are different or similar. This has been epitomized in the tectonic subdivision model of Zhao et al. (1998, 2001a, 2005) who discovered two Paleoproterozoic collisional orogens within the NCC, named the Trans-North China Orogen and the Khondalite Belt (Figure 1.4), of which the north-south-trending Trans-North China Orogen divides the basement of the NCC into the Eastern and Western Blocks, and the east-west-trending Khondalite Belt subdivides the Western Block into the Yinshan Block in the north and the Ordos Block in the south (Figure 1.4). Zhao et al. (2002b, 2005) proposed that the Khondalite Belt was a continent—continent collisional belt along which the Yinshan and Ordos blocks amalgamated to form the Western Block at ~1.95 Ga (see details in Chapter 4), and then the Western Block collided with a continental margin arc above an eastward subduction zone beneath the western margin of the Eastern Block to form the Trans-North China Orogen at ~1.85 Ga, leading to the final assembly of the NCC (see details in Chapter 4). In addition to the newly recognized Trans-North China Orogen and Khondalite Belt, there is another Paleoproterozoic tectonic belt located in the Eastern Block,

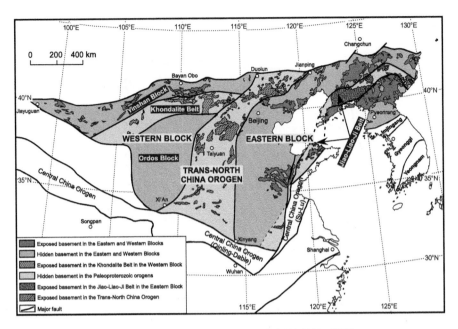

Figure 1.4 Tectonic subdivision of the NCC proposed by Zhao et al. (1998, 2001a, 2005).

called the Jiao-Liao-Ji Belt (Figure 1.4), which was recognized in the 1980s (Zhang and Yang, 1988). The tectonic nature of the Jiao-Liao-Ji Belt is still controversial, with some invoking arc-continent collision (Bai, 1993; Faure et al., 2004; Lu et al., 2006) and others involving the opening and closing of an intracontinental rift (Zhang and Yang, 1988; Li et al., 2004a,b, 2005, 2006, 2012; Luo et al., 2004, 2006, 2008; Li and Zhao, 2007). Most recently, Zhao et al. (2011a, 2012) interpreted the Jiao-Liao-Ji Belt as a Paleoproterozoic rift-and-collision belt within the Eastern Block, which encountered a rifting event in the period 2.2–1.9 Ga, leading to the opening of an incipient ocean that broke up the Eastern Block into the two small blocks, named the Longgang and Langrim blocks, which were reassembled through subduction and collision to form the Jiao-Liao-Ji Belt at ∼1.9 Ga (see details in Chapter 4).

Other researchers have also confirmed the existence of the Khondalite Belt, Trans-North China Orogen, and Jiao-Liao-Ji Belt in the NCC, though they assigned different names to them and proposed different models for the formation and evolution of these Paleoproterozoic orogens (Zhai and Peng, 2007; Faure et al., 2007; Trap et al., 2007, 2008, 2009a,b, 2011, 2012; Santosh, 2010). For example, Zhai and Peng (2007) also recognized three Paleoproterozoic tectonic belts in the eastern, central, and western parts of the NCC, named the Liaoji, Jinyu, and Fengzhen orogenic/mobile belts, respectively (Figure 1.5), which are spatially coincident with the Jiao-Liao-Ji Belt, Trans-North China Orogen, and Khondalite Belt of Zhao et al. (1998, 2001a, 2005), respectively. However, Zhai et al. (2010) suggested that the development of these Paleoproterozoic orogenic/mobile belts was not related to the assembly of the NCC, which they consider to have formed by amalgamation of seven microcontinental blocks along the 2.7 and ~2.5 Ga Neoarchean greenstone belts at ~2.5 Ga (Figure 1.3). Zhai et al. (2010) proposed that the formation and evolution of these Paleoproterozoic orogenic/mobile belts were involved in the initial rifting to form incipient oceanic basins followed by subduction and collision (rift basin closure) to form various high-pressure granulites and ultrahigh-temperature (UHT) rocks in these belts. Obviously, this model also advocates the involvement of subduction- and collision-related orogenic processes during the development of these Paleoproterozoic belts, though it argues that the NCC was initially cratonized to become a coherent craton at the end of Neoarchean (~2.5 Ga), before the formation of these Paleoproterozoic orogenic/mobile belts.

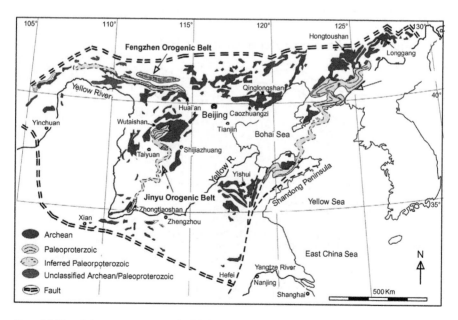

Figure 1.5 Three Paleoproterozoic orogenic/mobile belts of the NCC proposed by Zhai and Peng (2007).

Professor Timothy Kusky and his colleagues presented a series of tectonic models for the amalgamation and subsequent evolution of the NCC (Figure 1.6; Kusky and Li, 2003; Kusky et al., 2007; Kusky and Santosh, 2009; Li and Kusky, 2007; Kusky, 2011a; Deng et al., 2013). These models were established mainly upon two collisional orogens they claim to have recognized, named the Central Orogenic Belt and the Inner Mongolia–Northern Hebei Orogen (IMNHO) (Figure 1.6; Kusky and Li, 2003), of which the former is spatially coincident with the Trans-North China Orogen of Zhao et al. (2001a), though their boundaries are not completely the same, whereas the latter is bordered on the northern margin of the NCC (Figure 1.6). Initially, Kusky and Li (2003) proposed that the Central Orogenic Belt formed at ∼2.5 Ga when the Eastern and Western Blocks collided above a west-dipping subduction zone, whereas the 1400 km long, east-west-trending, IMNHO formed when an arc terrane collided with the northern margin of the NCC at ∼2.3 Ga, which was then converted to an Andean-style convergent margin in the period 2.20–1.85 Ga. Later, Kusky et al. (2007) modified their early model by proposing that in the period 1.92–1.85 Ga, the northern margin (IMNHO) of the NCC collided with part (South America?) of the Columbia supercontinent, and by

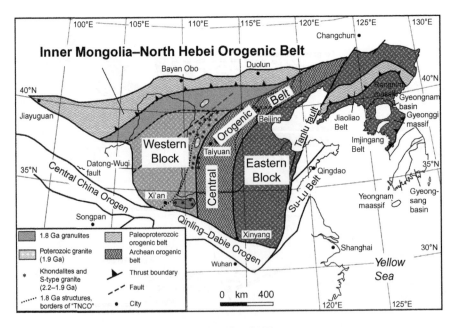

Figure 1.6 Tectonic subdivision of the NCC by Kusky and Li (2003).

1.8 Ga, the craton had switched into an extensional mode, breaking out of Columbia. Most recently, Kusky (2011a) and Deng et al. (2013) have made further modifications on the Kusky and Li (2003) model, especially for the development of the Central Orogenic Belt and the amalgamation of the Eastern and Western Blocks. They propose that in the period prior to 2.5 Ga, an intraoceanic arc terrane with a forearc assemblage developed above a west-dipping subduction zone within the ocean between the Eastern and Western Blocks, and at ∼2.5 Ga, the arc terrane collided with the western passive margin of the Eastern Block. Following the arc-continent collision at ∼2.5 Ga, an eastward-dipping subduction zone developed beneath the collided arc, which eventually caused the closure of the ocean basin, leading to the final collision of the Western Block with the arc-collision modified western margin of the Eastern Block. This new model implies that the final collision between the Eastern Block (united with an arc) and the Western Block occurred above an eastward subduction zone in the Paleoproterozoic, nearly close to the model proposed by Zhao et al. (2001a, 2005).

In the last few years, a French-China research team led by Prof. Michel Faure from University of Orleans has carried out extensive investigations on the Trans-North China Orogen and proposed a model different from others (Figure 1.7; Faure et al., 2007; Trap et al.,

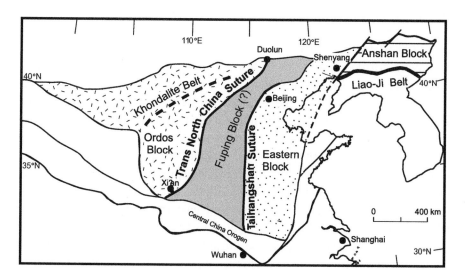

Figure 1.7 Tectonic subdivision of the NCC proposed by Faure et al. (2007).

2007, 2008, 2009a,b, 2011, 2012). The model postulates that there was an old continental block, named the Fuping Block, intervening between the Eastern and Western (Ordos) Blocks. According to Faure et al. (2007) and Trap et al. (2008), the Fuping Block was separated from the Eastern and Western Blocks by the Zanhuang and Lüliang oceans, respectively, and both of the oceans were westward subducted. At ~2.1 Ga, the westward-dipping subduction caused the closure of the Zanhuang Ocean at ~2.1 Ga, leading to the amalgamation of the Eastern and Fuping Blocks along the Taihangshan Suture, whereas the closure of the Lüliang Ocean took place at 1.9–1.8 Ga, leading to the final assembly of the Eastern and Western (Ordos) Blocks along the Trans-North China Suture (Figure 1.7; Faure et al., 2007; Trap et al., 2009a,b, 2011, 2012).

Prof. M. Santosh and his Chinese collaborators have done a great deal of work on the UHT metamorphism of the Khondalite Belt (Santosh et al., 2006, 2007a,b, 2008, 2009a,b, 2010, 2012, 2013), which they named "the Inner Mongolia Suture Zone" (Santosh, 2010). On the basis of overall synthesis of geological and geophysical data from both the Khondalite Belt and the Trans-North China Orogen, Santosh (2010) proposed a double-side subduction model for the Paleoproterozoic amalgamation of the NCC (Figure 1.8). By reinterpreting the seismic images of Chen et al. (2009), Santosh (2010) suggested that the present west-dipping shape of the lithosphere–asthenosphere boundary (LAB) across the Trans-North China Orogen formed by the westward subduction of the Eastern Block during Paleoproterozoic time, though Chen et al. (2009) interpreted the current structures of the LAB in the NCC as having formed during the Mesozoic to Cenozoic lithospheric thinning/destruction. Combining geological and geophysical data for the lithotectonic units within the Khondalite Belt (Inner Mongolia Suture Zone) and the Trans-North China Orogen, Santosh (2010) proposed that the Yinshan Block was oblique eastward subducted beneath the western (present-day) margin of the Yanliao (Eastern) Block and was southward subducted beneath the northern (present-day) margin of the Ordos Block. Therefore, the Ordos Block underwent the oppositely-verging subduction on its northern and eastern margins (Figure 1.8), which is considered to have contributed to the rapid amalgamation one of the major cratons on the globe during Paleoproterozoic time and its incorporation within the Columbia supercontinent (Santosh, 2010; Santosh et al., 2010).

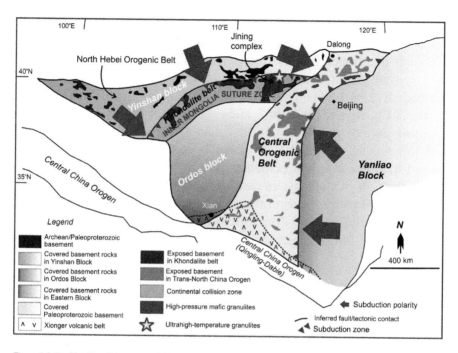

Figure 1.8 Double-side subduction model for the assembly of the NCC proposed by Santosh (2010).

1.3 SUMMARY

Although comprehensive geological and geophysical investigations have been carried out on the Precambrian basement of the NCC since the late 1990s, a broad agreement has not been reached on the tectonic subdivision and evolution of the craton; and as discussed above, the current tectonic models are largely mutually exclusive in terms of timing and tectonic processes of the collisional events that led to the amalgamation of the NCC. On the other hand, however, nearly all the models accept the subdivision of the NCC into the Eastern and Western Blocks which amalgamated along the Trans-North China Orogen (Jinyu Belt/Central Orogenic Belt) to form the coherent basement of the craton, though there are hot debates on the timing and tectonic processes of collision between the two blocks. In addition, most of the models concede the existence of a Paleoproterozoic orogenic belt within the Western Block, which has been given different names, including the Khondalite Belt (Zhao et al., 2005), the Fengzhen Belt (Zhai and Peng, 2007), and the Inner Mongolia Suture Zone (Santosh, 2010). There is no disagreement about the existence of the Jiao-Liao-Ji

Belt (also called the Liaoji Belt; Zhai and Peng, 2007) within the Eastern Block, but the existence of the IMNHO of Kusky and Li (2003) is questioned because the Inner Mongolian portion of this belt is a typical Neoarchean terrain (Yinshan Block) in which all rocks formed at 2.6–2.5 Ga and were metamorphosed at ∼2.5 Ga (Jian et al., 2012), whereas the north Hebei portion of this belt is the north-ernmost part of the Trans-North China Orogen (Zhao et al., 2012). For this reason, the "IMNHO" of Kusky and Li (2003) will not be dis-cussed in this book, which focuses on the Neoarchean accretion and amalgamation of the Eastern and Western Blocks and the formation and evolution of the Paleoproterozoic Khondalite Belt (Fengzhen Belt/ Inner Mongolia Suture Zone), Jiao-Liao-Ji Belt (Liaoji Belt), and Trans-North China Orogen (Jinyu Belt/Central Orogenic Belt).

CHAPTER 2

Archean Geology of the Eastern Block

2.1 INTRODUCTION

The Eastern Block is the epitome of Archean crust in the North China Craton (NCC) because most Archean rocks in the NCC are exposed in this block. The block consists of Archean to Paleoproterozoic basement overlain by Mesoproterozoic to Cenozoic unmetamorphosed cover. The Paleoproterozoic basement in the Eastern Block is restricted to the Jiao-Liao-Ji Belt (Figure 2.1), which will be discussed in Chapter 4. The Archean basement is exposed as high-grade gneissic complexes and low-grade granite-greenstone terranes in Luxi (LX), Yishui (YS), Qixia (QX), Miyun (MY), Eastern Hebei (EH), Jianping (JP), Fuxin-Suizhong (FX-SZ), Southern Liaoning (SL), Anshan-Benxi (AB), North Liaoning (NL), Southern Jilin (SL), and Langrim (LG) (Figure 2.1). The Hadean rocks have not been found in the Eastern Block. Considering that the 3.8 Ga zircons from the Anshan and Eastern Hebei areas have Hf isotopic compositions similar to those of chondrites, whereas the 3.6 Ga zircons from the Xinyang area in the southwesternmost part of the Eastern Block have lower Hf isotopic compositions, Wu et al. (2007) predicted that a >4.0 Ga Hadean crust may have existed in the Xinyang area, but not in the Anshan and Eastern Hebei areas. Most recently, however, Cui et al. (2013) discovered a Hadean zircon with a $^{207}Pb/^{206}Pb$ age of 4174 ± 48 Ma from the 2523 ± 12 Ma amphibolite in the Anshan area, which seems to support the existence of a Hadean crust in the Eastern Block. The Eoarchean (3.8–3.6 Ga) basement rocks in the Eastern Block are extremely scarce, only found in the Anshan area, represented by the ~3.8 Ga Chentaigou trondhjemitic gneisses (Figure 2.1; Jahn et al., 1987; Liu et al., 1992, 2007, 2008a; Song et al., 1996; Wan et al., 2012c), though large amounts of Eoarchean (3.88–3.55 Ga) detrital zircons have been found in the Caozhuangzi fuchsite quartzite and paragneiss in Eastern Hebei (Liu et al., 1992, 2007, 2013a; Wu et al., 2005a; Wilde et al., 2008; Nutman et al., 2011). The Paleo-Mesoarchean (3.6–2.8 Ga) basement rocks are also very rare, limited to the Anshan, Eastern Hebei, and Qixia areas, including the 3362–3342 Ma Chentaigou

Precambrian Evolution of the North China Craton. DOI: http://dx.doi.org/10.1016/B978-0-12-407227-5.00002-X

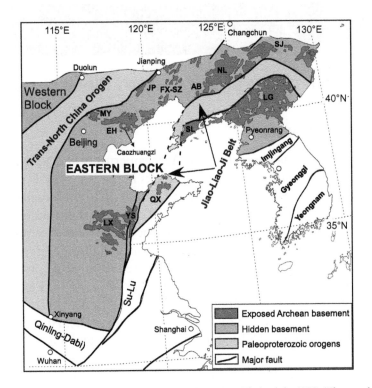

Figure 2.1 Distribution of Archean basement rocks in the Eastern Block of the NCC (Zhao et al., 2005). Abbreviations: AB, Anshan-Benxi; EH, Eastern Hebei; FX-SZ, Fuxin-Shuizhong; JP, Jianping; LG, Langrim; LX, Luxi; MY, Miyun; NL, Northern Liaoning; QX, Qixia; SJ, Southern Jilin; SL, Southern Liaoning; YS, Yishui.

supracrustal rocks, 3142 ± 7 Ma Lishan granites, 2994 ± 8 Ma Eastern Anshan granites, and 2962 ± 4 Ma Tiejiashan granites in the Anshan area (Liu et al., 1992, 2007, 2008a; Song et al., 1996); the 3.6–3.4 Ga Caozhuang supracrustals, the ~ 3.0 Ga Qianan supracrustal rocks, and ~ 2.98 Ga Yangyashan granites in Eastern Hebei (Jahn et al., 1987; Liu et al., 1992; Wilde et al., 2008); and 2892 ± 18 Ma supracrustals, 2906 ± 12 Ma tonalitic–trondhjemitic–granodioritic-granitic (TTG) gneisses, 2865 ± 18 Ma granitic gneisses in the Qixia area (Jahn et al., 2008; Wu et al., 2013a). These pre-Neoarchean rocks may have experienced multiple episodes of metamorphism and deformation between 3.8 and 2.5 Ga, but much of their petrographic and isotopic information on the early tectonothermal events has been obliterated by the last metamorphic event at ~ 2.5 Ga. Therefore, the tectonic settings, spatial scope, and tectonothermal evolution of the pre-Neoarchean crust in the Eastern Block of the NCC remain unknown.

The dominant Archean lithologies in the Eastern Block are Neoarchean (2.75–2.50 Ga) rocks, which are widespread throughout the block, making up more than 95% of total exposure of the Archean basement (Zhao et al., 2001a). They consist predominantly of TTG gneisses with minor supracrustal rocks (low- to high-grade greenstone sequences) ranging in age from 2.75 to 2.50 Ga, which are intruded by ∼2.5 Ga massive syntectonic granites/charnockites and experienced greenschist to granulite-facies metamorphism at ∼2.5 Ga (Pidgeon, 1980; Jahn and Zhang, 1984; Zhai et al., 1985, 1990, 2003; Jahn et al., 1988; Shen and Qian, 1995; Shen et al., 1992; Kröner et al., 1998; Zhao et al., 1998). Like Neoarchean high-grade gnessic complexes and low-grade granite-greenstone terranes in other cratonic blocks (e.g., South India, Yilgarn, Zambabwe, Baltica, Siberia, Superior, and Amazonia) in the world, the Neoarchean lithologies in the Eastern Block of the NCC possess some unique features in terms of their rock associations, metamorphism, and structural patterns, which are distinctly different from those of Phanerozoic terranes that developed under plate tectonic regimes. This raises a few long-debated questions in the geological community:

1. Whether or not did Archean high-grade gneissic complexes and low-grade greenstone terranes develop under the regime of plate tectonics?
2. If not, what was the main tectonic mechanism/regime that governed the formation and evolution of Archean crust?
3. When did plate tectonics begin on Earth (Condie and Kröner, 2008)?

These questions have long been nagging to geologists. In this chapter, we attempt to explore some of these questions through characterizing the rock association, geochemical composition, metamorphic evolution, and structural pattern of the Neoarchean crust in the Eastern Block.

2.2 A HADEAN (> 3.85 GA) CRUST IN THE EASTERN BLOCK

As the first geologic eon before the Archean, the Hadean started at Earth's formation about 4.6 Ga, and ended roughly 4.0–3.8 billion years ago, though the latter date varies according to different sources. Although the International Commission on Stratigraphy (ICS) has recently stipulated 4.0 Ga as the boundary between the Hadean and Archean, in this book I prefer to use 3.85 Ga as the end-time of the

Hadean as the Late Heavy Bombardment had not ended until ~ 3.85 Ga (Goldblatt et al., 2010). If so, available data support the existence of a Hadean crust in the Eastern Block, though its distribution extent remains unknown. The existence of such a Hadean crust in the Eastern Block is mainly evidenced from the zircon Hf model ages of Eoarchean to Mesoarchean rocks in the Anshan and Eastern Hebei areas. For example, Liu et al. (2008a) and Wu et al. (2008, 2009) carried out a preliminary investigation of the hafnium systematics of the ancient zircons in the Anshan area, and the results show that an adequate amounts of Eoarchean and Paleoarchean zircons from the Baijiafen, Dongshan, and Shengousi complexes have Hf model ages of >4.0 Ga, despite most at 3.7–4.0 Ga. This suggests that there was a Hadean crust in the Anshan area before the formation of Eoarchean and Paleoarchean rocks in the area. This is strongly supported by a recent discovery of a zircon with a $^{207}Pb/^{206}Pb$ age of 4174 ± 48 Ma from the 2523 ± 12 Ma amphibolite in the Anshan area (Cui et al., 2013). In addition to the Anshan area, the Eastern Hebei is regarded as another potential area with a Hadean proto-crust, which was most likely a Hadean continental crust dominated by felsic composition as the Caozhuang fuchsite quartzite contains a large amount of 3.55–3.85 Ga detrital zircons with $T_{DM}(Hf)$ model ages of >3.85 Ga (Wu et al., 2005a). Although Wu et al. (2005a) concluded that there was no large-scale pre-3.8 Ga crust in Eastern Hebei based on the $T_{DM}(Hf)$ model ages of Eoarchean zircons from the Caozhuang fuchsite quartzite, Wilde et al. (2008) obtained a SHRIMP $^{207}Pb/^{206}Pb$ age of 3860 ± 3 Ma for one zircon grain, and seven zircon grains record a concordia $^{207}Pb/^{206}Pb$ age of 3832 ± 4 Ma. Later, Nutman et al. (2011) also obtained a SHRIMP $^{207}Pb/^{206}Pb$ age of 3881 ± 38 Ma for a zircon from the fuchsite quartzite. These data imply that it was most likely that there was a Hadean crust in Eastern Hebei. In the Xinyang area that is located in the southwestern margin of the Eastern Block, Zheng et al. (2004a) and Liu et al. (2007) recognized a number of Paleoarchean (3650–3670 Ma) zircons with $T_{DM}(Hf)$ model ages of 4.0–3.9 Ga, implying a Hadean crust in the region.

2.3 EOARCHEAN (3.85–3.6 GA) ROCKS

The Eoarchean (3.8–3.6 Ga) rocks in the Eastern Block have been found only in the Anshan area (Figure 2.2), in which the oldest rocks are 3.8 Ga trondhjemite-dominated gneisses, represented by the

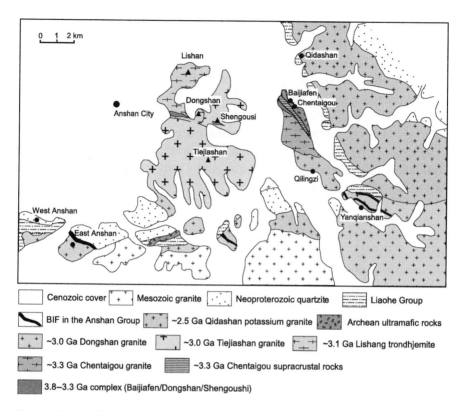

Figure 2.2 Regional distribution of Eoarchean to Neoarchean basement rocks in the Anshan area. After Wu et al. (1998) and Liu et al. (2008a).

Dongshan banded trondhjemitic gneisses with weighted mean SHRIMP $^{207}Pb/^{206}Pb$ zircon ages up to 3811 ± 4 Ma (Liu et al., 2007, 2008a), the Baijiafen mylonitized trondhjemitic gneisses with weighted mean SHRIMP $^{207}Pb/^{206}Pb$ zircon ages up to 3804 ± 5 Ma (Liu et al., 1992; Song et al., 1996), the Dongshan quartz dioritic gneiss with a weighted mean SHRIMP $^{207}Pb/^{206}Pb$ zircon age of 3792 ± 4 Ma (Wan et al., 2005a), and the Shengousi banded tronhjemitic gneisses of 3773 ± 6 Ma (Liu et al., 2007, 2008a). These ~ 3.8 Ga trondhjemite and dioritic gneisses are enclosed within relatively younger phases of Eoarchean trondhjemites that were emplaced at 3680 ± 19 Ma in Dongshan and cut by a 3620 ± 23 Ma trondhjemitic vein in Baijiafen (Liu et al., 2008a). Associated with these Eoarchean plutons are minor Eoarchean supracrustal rocks, of which a biotite schist contains a zircon population with a weighted mean $^{207}Pb/^{206}Pb$ age of 3723 ± 17 Ma and itself is cut by 3620 ± 23 Ma trondhjemitic dyke, implying that its sedimentary protoliths must have been deposited at some time prior to 3.62 Ga (Liu et al., 2008a). The T_{DM}(Hf) model ages of zircons from these Eoarchean rocks in the Anshan area range from 3256 to 4264 Ma, with a most prominent peak at 3.9 Ga, indicating the presence of a crust older than the 3.8 Ga trondhjemites (Wu et al., 2008). All these data further confirm the existence of a Hadean–Eoarchean crust in the Anshan area, though the Eoarchean rocks crop out only in a small area ($<20\ km^2$) and the real extent of Eoarchean crust presented by these rocks still remains controversial (Wu et al., 2008, 2009).

Although Eoarchean rocks are only exposed in the Anshan area, Eoarchean zircons have been reported from other parts of the Eastern Block, especially in Eastern Hebei where large amounts of Eoarchean zircons are reported from the Caozhuang Group, which crops out in the Huangbaiyu and Xingshan areas (Figure 2.3). Using the SHRIMP ion microprobe technique, Liu et al. (1992) dated a fuchsite quartzite

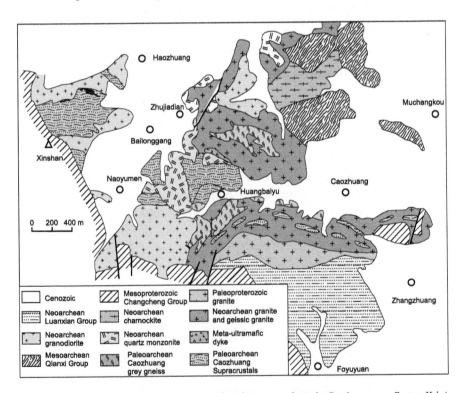

Figure 2.3 Regional distribution of Early to Late Archean basement rocks in the Caozhuang area, Eastern Hebei. After Wu et al. (1998).

sample from the Caozhuang Group, and 81 analyses on 61 zircon grains yielded $^{207}Pb/^{206}Pb$ apparent ages ranging from 3.55 to 3.85 Ga, of which most zircons show oscillatory zoning, interpreted as detrital igneous grains, though one zircon grain with a $^{207}Pb/^{206}Pb$ apparent age of 3.55 Ga has a Th/U ratio of 0.02, interpreted to be of metamorphic origin. Liu et al. (2007) interpreted the age of 3.55 Ga as the timing of a metamorphic event occurring in the source region, whereas the sedimentary protolith of the fuchsite quartzite is considered to have been deposited at some time after 3.55 Ga (Nutman et al., 2011). Using the LA-ICP-MS technique, Wu et al. (2005a) analyzed another sample of fuchsite quartzite from a nearby outcrop at Huangbaiyu, and obtained a similar spread of ages, but their $T_{DM}(Hf)$ model ages and $\varepsilon Hf_{(t)}$ data led to a conclusion that there was no large-scale pre-3.8 Ga crust in this area. Most recently, using the SHRIMP technique, Liu et al. (2013a) dated garnet-biotite and hornblende paragneisses that are associated with the fuchsite quartzite in the Caozhuang area and revealed a number of Eoarchean zircons from these rocks, further confirming the existence of the Eoarchean crust in Eastern Hebei. Eoarchean zircons of ∼3.6 Ga have also been revealed from felsic granulite xenoliths in the Mesozoic volcanics in Xinyang, which is situated at the southwestern margin of the NCC (Zheng et al., 2004a). This suggests that the Eoarchean crust is not only restricted to the eastern part of the Eastern Block but may also exist in the western part of the block (Zheng et al., 2004a).

2.4 PALEOARCHEAN (3.6–3.2 GA) ROCKS

Recent SHRIMP and laser ablation-induced coupled plasma-mass spectrometry (LA-ICP-MS) zircon data indicate that like the Eoarchean (3.8–3.6 Ga) lithologies in the Eastern Block, the Paleoarchean (3.6–3.2 Ga) rocks are also only exposed in the Anshan and Eastern Hebei, represented by the Shengousi trondhjemite, Chentaigou granite, and Chentaigou supracrustal rocks in the Anshan area (Figure 2.2; Song et al., 1996; Wu et al., 1998) and the Caozhuang amphibolites in Eastern Hebei (Figure 2.3; Huang et al., 1986; Jahn et al., 1987; Qiao et al., 1987; Wu et al., 1998).

In the Anshan area, the Shengousi Complex is situated between the Baijiafen and Dongshan complexes and consists of fine-grained gneissic trondhjemites as strips and lenses within granitic and pegmatitic migmatites, and coarse-grained banded trondhjemites that are in concordant contact with foliated pegmatitic monzogranite without compositional banding (Wan et al., 2012c). The fine-grained gneissic trondhjemite and coarse-grained banded trondhjemite yielded SHRIMP ^{207}Pb/^{206}Pb zircon ages of 3454 ± 8 and 3448 ± 9 Ma, respectively, interpreted as their crystallization ages (Wan et al., 2012c). These trondhjemitic gneisses are considered as another phase (Phase III) of trondhjemitic magmatism following the early Eoarchean Phase I (3.81−3.77 Ga) and late Eoarchean Phase II (3.62−3.57 Ga) trondhjemitic magmatism (Liu et al., 1992, 2007, 2008a; Song et al., 1996; Wan et al., 2005a, 2012c; Wu et al., 2008). The Chentaigou supracrustal rocks are composed of amphibolites, biotite-plagioclase gneiss, leptynite, quartzite, and minor calc-silicates (Wu et al., 1998), of which the leptynite yielded a SHRIMP ^{207}Pb/^{206}Pb zircon age of 3362 ± 5 Ma and a granitic dyke intruding the Chentaigou supracrustal rocks gave a SHRIMP ^{207}Pb/^{206}Pb zircon age of 3342 ± 10 Ma (Song et al., 1996), suggesting that the Chentaigou supracrustal rocks must have formed at some time between 3362 and 3342 Ma (Wu et al., 1998). The Chentaigou granite is rather homogeneous with an augen texture derived from the deformation of K-feldspar megacrysts (Song et al., 1996), and igneous zircons from the granite yielded a SHRIMP ^{207}Pb/^{206}Pb zircon age of 3306 ± 13 Ma (Song et al., 1996). The late Paleoarchean granitoid rocks are also exposed in the Dongshan Complex, represented by the Dongshan dioritic gneiss and the Dongshan trondhjemitic gneiss, which were dated at 3303 ± 5 and 3305 ± 7 Ma, respectively (Wu et al., 2008); the latter is the youngest phase (Phase IV) of trondhjemite in the Anshan area.

In Eastern Hebei, Paleoarchean rocks consist of the Caozhuang supracrustal rocks (Figure 2.3; Zhao and Zhai, 2013), which comprise fuchsite quartzite, garnet quartzite, garnet-biotite paragneiss, sillimanite-biotite gneiss, calc-silicate rocks, marble, banded iron formation (BIF), and minor amphibolite, of which the amphibolites yielded Sm−Nd whole-rock isochron ages of 3500 ± 80 (Huang et al., 1986), 3470 ± 107 (Jahn et al., 1987), and 3561 ± 15 Ma (Qiao et al., 1987).

2.5 MESOARCHEAN CRUST (3.2−2.8 GA) ROCKS

Previous zircon ages, mostly obtained by using the conventional multi-grain zircon dating techniques, led to a conclusion that the

Mesoarchean supracrustals and granitoids were widely distributed in the Eastern Block, especially in Eastern Hebei, Anshan, Northern Liaoning, Southern Jilin, Western Shandong (Luxi), and Eastern Shandong (Ma and Wu, 1981; Dai et al., 1990; Bai, 1993; Bai and Dai, 1998; Wu et al., 1998). However, this conclusion has been denied by recent SHRIMP and LA-ICP-MS zircon data, which indicate that like the Eoarchean and Paleoarchean lithologies in the Eastern Block, the Mesoarchean (3.2−2.8 Ga) rocks are also very scarce and only exposed in the Anshan, Eastern Hebei, and Qixia areas, represented by the Lishan, Eastern Anshan, and Tiejiashan granites in the Anshan area (Figure 2.2; Song et al., 1996; Wu et al., 1998); the Qianan supracrustal rocks and Yangyashan granites in Eastern Hebei (Figure 2.3; Huang et al., 1986; Jahn et al., 1987; Qiao et al., 1987; Wu et al., 1998); and the Huangyadi TTG gneisses and Xiaoshibapan supracrustals in the Qixia area (Jahn et al., 2008; Wu et al., 2013a). In the Anshan area, the Mesoarchean granites mainly occur in the Lishan and Tiejiashan areas where they occur as a dome. The Lishan granite is generally fine grained in texture and relatively massive in structure, without pronounced pegmatitic banding (Song et al., 1996). It contains abundant enclaves of amphibolite, amphibolitic gneiss, and biotite schist. A Lishan granite sample collected in the Lishan Park yielded a SHRIMP ^{207}Pb/^{206}Pb zircon age of 3142 ± 7 Ma, interpreted as the emplacement age of the Lishan granite (Song et al., 1996), which is the oldest phase of the Mesoarchean magmatism in the Eastern Block. The Tiejiashan granite is weakly porphyritic medium-grained biotite K-rich granite, three samples of which yielded SHRIMP ^{207}Pb/^{206}Pb zircon age of 2992 ± 10, 2983 ± 10, and 2962 ± 4 Ma, interpreted as the crystallization age of the granite (Song et al., 1996; Wan et al., 2007). In addition to the Lishan−Tiejiashan area, the Mesoarchean granites have also been recognized from the West and East Anshan BIF mining areas, of which the West and East Anshan granites yielded a SHRIMP ^{207}Pb/^{206}Pb zircon ages of 3001 ± 8 and 2994 ± 8 Ma, respectively (Song et al., 1996).

In Eastern Hebei, Mesoarchean rocks consist of the Qianan supracrustal rocks and Yangyashan granites (Figure 2.3; Zhao and Zhai, 2013). The Qianan supracrustal rocks are exposed in the Shuichang-Songting area, with lithologies similar to those of the Caozhuang supracrustals, including amphibolites, mafic granulites, leptynites, sillimanite-garnet paragneisses, biotite-plagioclase paragneisses, BIF, quartzites, and minor calc-silicates (Wu et al., 1998), of which the BIFs are intruded by the Yangyashan orthogneiss that was dated (U−Pb zircon) at 2980 ± 8 Ma, defining a minimum depositional age of the Qianan supracrustal rocks (Wu et al., 1998).

In the Qixia area, the Mesoarchean Huangyadi TTG gneisses and minor supracrustals are enclosed within the Neoarchean TTG gneisses and supracrustal rocks, called the Jiaodong Group (Figure 2.4; Jahn et al., 2008). In the field, the Mesoarchean and Neoarchean TTG gneisses and supracrustals are difficult to be recognized from each

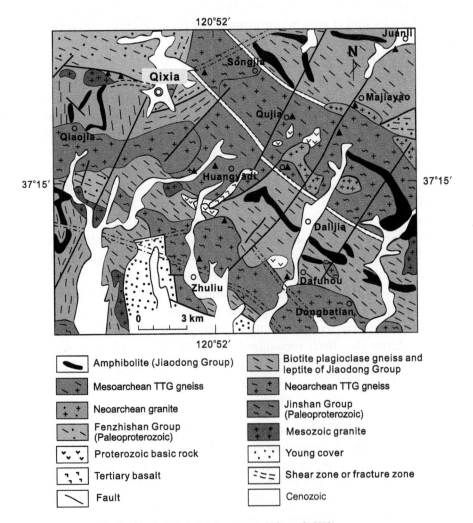

Figure 2.4 Geological map of the Qixia area, Jiaodong terrane (Jahn et al., 2008).

other. As shown in Figure 2.4, Jahn et al. (2008) mapped out a small Neoarchean TTG (grey) gneiss terrain associated with supracrustal rocks in the Huangyadi area of which a grey gneiss sample yielded a SHRIMP ^{207}Pb/^{206}Pb zircon age of 2906 ± 12 Ma and a fine-grained biotite leptite (paragneiss) sample gave a SHRIMP ^{207}Pb/^{206}Pb zircon age of 2892 ± 18. However, not all TTG gneisses in the Huangyadi area are Mesoarchean in age, and most of them are of Neoarchean ages (Wu et al., 2013a). On the other hand, Mesoarchean TTG gneisses are not only restricted to the Huangyadi area. For example, a granodioritic gneiss sample collected near Dachai Village, 21 km southeast of Qixia City, yielded a LA-ICP-MS age of 2865 ± 18 Ma, interpreted as the crystallization age of magmatism (Wu et al., 2013a). In addition, a supracrustal rock (medium-grained biotite-plagioclase gneiss) collected on an outcrop north of Daliujia Village yielded two age populations, of which the younger group gives apparent ^{207}Pb/^{206}Pb ages between 2804 and 2724 Ma and the older group gives apparent ^{207}Pb/^{206}Pb ages of 2916−2821 Ma (Wu et al., 2013a). All these data confirm the existence of Mesoarchean crust in the Qixia area in Eastern Shandong.

2.6 NEOARCHEAN (2.8−2.5 GA) ROCKS

According to available zircon age data, the Neoarchean crust of the Eastern Block can be further subdivided into the 2.75−2.65 and 2.55−2.50 Ga rock associations, of which the former is only exposed in local areas (e.g., Luxi and Qixia), whereas the latter makes up more than 90% of total exposure of the Archean basement in the block. The two Neoarchean rock associations display distinct lithological, meta-morphic, structural, and geochemical features.

2.6.1 2.75−2.65 Ga Rock Associations

In the Eastern Block, the 2.75−2.65 Ga rocks only occur in the Luxi Complex in Western Shandong and the Qixia Complex in Eastern Shandong (Jahn et al., 1988, 2008; Wan et al., 2011a,b; Wu et al., 2013a). In Western Shandong, the Luxi Complex is also called the Luxi granite-greenstone terrane that includes both the 2.75−2.65 and 2.55−2.50 Ga rock associations, of which the former are mainly exposed along a NW-SE-trending belt surrounded by the 2.55−2.50 Ga TTG gneisses and ~2.5 Ga syntectonic granites (Figure 2.5; Wan et al., 2011a, 2012a). The 2.75−2.65 Ga rock

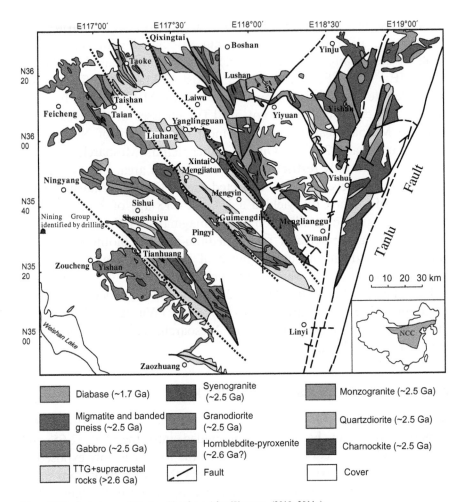

Figure 2.5 Geological map of Western Shandong. After Wan et a. (2010, 2011a).

associations in the Luxi granite-greenstone belt consist of TTG gneisses and greenstones, of which the latter occur as lenses and sheets within TTG gneisses and are traditionally called the Taishan "Group" in the Chinese literature (Cheng and Xu, 1991), which is further subdivided into the Yanlingguan, Shancaoyu, and Liuhang "Formations," though their original lithostratigraphic relationships are unclear (Wan et al., 2011a). Conventionally, it is considered that the Liuhang and Yanlingguan "Formations" are opposing limbs of a syncline, older than the Shancaoyu "Formation" that occupies the center of the syncline (Cao, 1996). The Yanlingguan "Formation" is a volcanic-rich unit that is transitional from a basal ultramafic- and mafic-rich assemblage, through a middle mafic-intermediate volcanic rock suite with minor BIFs, to an upper sedimentary-rich sequence. The basal ultramafic-rich assemblage includes ~400 m thick komatiitic rocks that are composed of serpentinized peridotite, tremolite-talc schist, and chlorite-actinolite schist, some of which preserve typical spinifex textures with plate-like crystals of 3–10 cm long serpentinized olivine (Figure 2.6). These komatiites are geochemically similar to other komatiites in the world and are interpreted to have resulted from mantle plume activities (Polat et al., 2006; Cheng and Kusky, 2007). Metabasaltic rocks and felsic gneisses from the Yanlingguan "Formation" yielded Rb–Sr and Sm–Nd whole-rock isochron ages of 2767 ± 45 and 2740 ± 74 Ma, respectively (Jahn et al., 1988), which are interpreted as the time of basaltic volcanism in the Yanlingguan "Formation." This is confirmed by SHRIMP ^{207}Pb/^{206}Pb zircon ages of 2747 ± 7 Ma for fine-grained hornblende-biotite gneiss from the Yanlingguan "Formation" and 2740 ± 6 Ma for a weak gneissic quartz diorite dyke cutting the ultramafic rocks of the Yanlingguan Formation (Table 2.1; Wan et al., 2011a). The Shancaoyu "Formation" of the Taishan "Group" is dominated by meta-greywackes, interlayered with minor mafic to felsic volcanics, metamorphosed from greenschist- to lower amphibolite-facies. The Liuhang "Formation" is lithologically similar to the Yanlingguan "Formation" but contains more sedimentary rocks. A fine-grained biotite gneiss (biotite-rich leptynite) sample and a mylonitized fine-grained biotite

Figure 2.6 Komatiites with spinifex textures from the Luxi granite-greenstone terrane.

Table 2.1 SHRIMP and LA-ICP-MS Zircon Ages for the 2.75–2.65 Ga Rock Associations in the Eastern Block

Sample No.	Rock type	Location/Pluton Name/Formation	Igneous/Protolithic Age (Ma)	Metamorphic Age (Ma)	Method	Sources
Luxi (Western Shandong)						
M08	Garnet quartzite	Mengjiatun F.	2717 ± 33	2616 ± 19	SHRIMP	Du et al. (2003)
SD0513	Garnet quartzite	Mengjiatun F.	2719	2640 ± 48	SHRIMP	Lu et al. (2008a)
D242	St–Grt mica schist	Mengjiatun F.	2742 ± 23	2642 ± 23	SHRIMP	Du et al. (2003)
M06	Amphibolite	Mengjiatun F.		2609 ± 12	SHRIMP	Du et al. (2003)
M03	Banded biotite gneiss	Mengjiatun F.	2695 ± 14	2624 ± 11	SHRIMP	Du et al. (2003)
S0701	Hbl–Bi leptynite	Yanglingguan F.	2747 ± 7		SHRIMP	Wan et al. (2011a)
S0725	Biotite leptynite	Liuhang F.	2739 ± 16	2606 ± 18	SHRIMP	Wan et al. (2011a)
S0721	Biotite leptynite	Liuhang F.	2703 ± 6		SHRIMP	Wan et al. (2011a)
SD0613-2	Quartz dioritic gneiss	Liuhang	2741 ± 47		LA-ICP-MS	Lu et al. (2008a)
SD0613-2	Quartz dioritic gneiss	Liuhang	2727 ± 37		SHRIMP	Lu et al. (2008a)
SY0311	Quartz diorite	E. Yanglingguan	2740 ± 6		SHRIMP	Wan et al. (2011a)
SY0336	Gneissic trondhjemite	Wangfushan	2711 ± 10		SHRIMP	Wan et al. (2011a)
S0503	Gneissic Tonalite	Wangfushan	2714 ± 13	2663–2644	SHRIMP	Wan et al. (2011a)
S0741	Gneissic granodiorite	Wangfushan	2712 ± 7		SHRIMP	Wan et al. (2011a)
S0732	Gneissic Trondhjemite	Duozhuang	2707 ± 9		SHRIMP	Wan et al. (2011a)

Sample	Rock type	Location	Age	Age	Method	Reference
SD0601	Amphibolite	Yanlingguan	2678 ± 26		LA-ICP-MS	Lu et al. (2008a)
SD0602	Anatectic felsic dyke	Yanlingguan	2663 ± 16	2663 ± 16	LA-ICP-MS	Lu et al. (2008a)
SD0604	Tonalite	Wangfushan	2637 ± 22		LA-ICP-MS	Lu et al. (2008a)
SD0512	Tonalite	Liuhang	2609 ± 19		LA-ICP-MS	Lu et al. (2008a)
SD0503	Trondhjemite	Liuhang	2611 ± 19		LA-ICP-MS	Lu et al. (2008a)
TS09100	Trondhjemitic dyke	Qixingtai	2706 ± 9		SHRIMP	Wang et al. (2013a)
JN0742	Tonalite	Taishan	2691 ± 7		LA-ICP-MS	Jiang et al. (2010)
Qixia (Eastern Shandong)						
10SD18-1	Granodioritic gneiss	Nanfojia Village	2712 ± 12	2503 ± 11	LA-ICP-MS	Wu et al. (2013b)
10SD21-1	Granodioritic gneiss	Huangyadi Village	2745 ± 12		LA-ICP-MS	Wu et al. (2013b)
10SD20-1	Tonalitic gneiss	Hedongya Village	2710 ± 14		LA-ICP-MS	Wu et al. (2013b)
CF92-34	Banded TTG gneiss	Dafuhou Village	2707 ± 4		SHRIMP	Jahn et al. (2008)
S0123-1	TTG gneiss	Zhuliu Village	2726 ± 12	2491 ± 8/2499 ± 18	SHRIMP	Jahn et al. (2008)
S0129-1	Mylonitized TTG gneiss	Majiayao Village	2718 ± 128	2489 ± 8/2486 ± 8	SHRIMP	Jahn et al. (2008)

gneiss sample from the Liuhang "Formation" yielded SHRIMP $^{207}Pb/^{206}Pb$ zircon ages of 2739 ± 16 and 2703 ± 6 Ma, respectively, interpreted as their rock-forming ages (Table 2.1; Wan et al., 2011a), which is further confirmed by a SHRIMP $^{207}Pb/^{206}Pb$ zircon age of 2706 ± 9 Ma for a trondhjemitic dike cutting the base of the pillow basalts of the Liuhang "Formation" (Table 2.1; Wang et al., 2013a). Associated with the Taishan "Group" are TTG suites, of which quartz diorite, gneissic tonalite, gneissic trondhjemite, and granodiorite yielded SHRIMP $^{207}Pb/^{206}Pb$ zircon ages of 2740 ± 6, 2714 ± 13, 2711 ± 10, and 2712 ± 7 Ma, respectively (Table 2.1; Wan et al., 2011a). Some zircons from the TTG gneisses possess narrow ($25-30$ μm) metamorphic overgrowth rims that produced $^{207}Pb/^{206}Pb$ ages ranging between 2663 and 2644 Ma, interpreted as the metamorphic ages (Wan et al., 2011a). This is consistent with the metamorphic zircon age of $2640-2620$ Ma obtained for the Mengjiatun "Formation" that is considered to be the equivalent of the Taishan "Group" (Table 2.1; Du et al., 2003; Lu et al., 2008a). These data indicate that like early Neoarchean rock assemblages in many other cratons (e.g., Yilgarn, Superior, Zambabwe, Baltica, and Amazonia), the $2.75-2.65$ Ga rock associations in the Eastern Block also experienced regional metamorphism at ~ 2.65 Ga (Wan et al., 2011a).

It deserves mentioning that recently, Wan et al. (2012a) reported that the Shancaoyu "Formation" and the minor part of the Yanglinguan and Liuhang "Formations" were deposited in the late Neoarchean ($2.55-2.52$ Ga), not at $2.75-2.65$ Ga as previously considered, indicating that not all lithostratigraphic units in the Taishan "Group" developed during early Neoarchean time, though there is no doubt that the majority of the Taishan "Group" formed at $2.75-2.65$ Ga as supported by zircon ages (Table 2.1; Lu et al., 2008a; Wan et al., 2011a; Wang et al., 2013a).

In Eastern Shandong, the Qixia Complex consists of Archean supracrustal rocks (metavolcanic and metasedimentary rocks), Archean TTG gneisses, Paleoproterozoic metasedimentary sequences (Jingshan and Fenzishan groups), and miscellaneous rock types of younger ages (Figure 2.4; Jahn et al., 2008; Wan et al., 2011b; Wu et al., 2013a). The Archean supracrustal rocks, traditionally named the Jiaodong Group in the Chinese literature, occur as boudins or lenses sparsely enclosed in the Neoarchean TTG gneisses, but in the vicinity of Huangyadi Village (Figure 2.4), they crop out on a large scale, locally called the Huangyadi Group (Jahn et al., 2008). The major rocks of the Huangyadi Group are biotite leptite, biotite-plagioclase gneiss, and amphibolites metamorphosed in amphibolite facies, of which the biotite leptite is considered to be meta-dacite (Jahn et al., 2008), whereas the amphibolites were metamorphosed basalts. Except minor Mesoarchean supracrustal rocks (2892 ± 18 Ma) and TTG gneisses (2906 ± 12 Ma), most TTG gneisses from the Qixia Complex are formed between 2.70 and 2.75 Ga (Table 2.1; Jahn et al., 2008; Wu et al., 2013a). Unlike the ∼2.75 Ga TTG gneisses in the Luxi Complex that underwent a metamorphic event ∼2.65 Ga, the TTG gneisses in the Qixia Complex only preserve the records of the ∼2.5 and 1.90−1.85 Ga metamorphic events (Table 2.1; Jahn et al., 2008; Wu et al., 2013a). A possible explanation is that the record of the early ∼2.65 Ga tectonothermal event in these rocks was completely obliterated by the later ∼2.5 and 1.90−1.85 Ga metamorphic events.

2.6.2 2.55−2.50 Ga Rock Associations

Recent developments in microanalysis, including secondary ion mass spectrometry (SIMS) and LA-ICP-MS, combined with the detailed imaging of internal structures by cathodoluminescence (CL) or back-scattered electrons (BSE), enable to precisely and accurately date magmatic and metamorphic events on single zircon grains. In the last decade, researchers have applied such zircon microanalysis techniques (e.g., SHRIMP, CAMECA, and LA-ICP-MS) to date the major lithologies of the Eastern Block of the NCC and obtained large amounts of new data (Shen et al., 2004, 2007; Wan et al., 2005b, 2012a,b, 2013; Geng et al., 2006, 2010, 2012; Lu et al., 2008a; Yang et al., 2008; Zhao et al., 2008c, 2009a, 2013; Grant et al., 2009; Zheng et al., 2009b; Li et al., 2010b; Liu et al., 2011a,b, 2013b; Nutman et al., 2011; Lü et al., 2012; Cui et al., 2013; Meng et al., 2013a,b; Peng et al., 2013a,b; Wu et al., 2013a,b,c). These new data, combined with those ages previously produced by the conventional zircon dating methods, confirm that more than 90% of the exposed Archean lithologies in the Eastern Block were formed in the period 2.55−2.50 Ga. Table 2.2 lists all available zircon ages for 2.55−2.50 Ga rock associations from different metamorphic complexes in the Eastern Block.

The 2.55−2.50 Ga rock associations in the Eastern Block include 2.55−2.50 Ga TTG gneisses, 2.6−2.5 Ga ultramafic to mafic igneous

Table 2.2 Zircon Ages for the 2.55–2.50 Ga Rock Associations in the Eastern Block

Sample No.	Rock Type	Location	Igneous and Protolithic Age (Ma)	Metamorphic Age (Ma)	Method	Sources
Miyun						
FW04-120	TTG gneiss	SE Miyun	2521 ± 14		LA-ICP-MS	Yang et al. (2005)
Eastern Hebei						
FW04-28	Dioritic gneiss	Jiekouling	2528 ± 3	2501 ± 8	LA-ICP-MS	Yang et al. (2008)
FW04-42	Granodioritic gneiss	Qinhuangdao	2522 ± 5	2497 ± 4	LA-ICP-MS	Yang et al. (2008)
FW04-51	Granodioritic gneiss	Qinhuangdao	2527 ± 5		LA-ICP-MS	Yang et al. (2008)
FW04-54	Kf-granitic gneiss	Qinhuangdao	2523 ± 6	2451 ± 6	LA-ICP-MS	Yang et al. (2008)
FW04-84	Granodioritic gneiss	Anziling	2524 ± 8	2490 ± 4	LA-ICP-MS	Yang et al. (2008)
FW04-85	Tonalitic gneiss	Anziling	2522 ± 8	2501 ± 8	LA-ICP-MS	Yang et al. (2008)
04QA08	Olivine gabbro dyke	Qian'an	2516 ± 26		SHRIMP	Li et al. (2010b)
04QA08	Syenite dyke	Qian'an	2504 ± 11		SHRIMP	Li et al. (2010b)
J08/12	Monzogranite	Beidaihe	2512 ± 12		SHRIMP	Nutman et al. (2011)
J08/15	Tonalitic gneiss	Beidaihe	2546 ± 6		SHRIMP	Nutman et al. (2011)
J08/16	Monzogranite	Beidaihe	2525 ± 10		SHRIMP	Nutman et al. (2011)
J08/18	Mafic granitoid	Qinhuangdao	2531 ± 3		SHRIMP	Nutman et al. (2011)
J08/05	High-grade TTG gneiss	Santunying	2525 ± 20		SHRIMP	Nutman et al. (2011)
J08/09	Granulite-facies pegmatite	Santunying	2525 ± 20	2525 ± 20	SHRIMP	Nutman et al. (2011)
J08/10	High-grade TTG gneiss	Santunying	2537 ± 9	2498 ± 6	SHRIMP	Nutman et al. (2011)
J06/10	High-grade gneiss	Caozhuangzi	2548 ± 7	2506 ± 6	SHRIMP	Nutman et al. (2011)
J91/11	Tonalitic gneiss	Caozhuangzi	2548 ± 13		SHRIMP	Nutman et al. (2011)

J00/33	Leptynite (volcanic)	Caozhuangzi	2534 ± 8		SHRIMP	Nutman et al. (2011)
J00/31	Monzonite	Caozhuangzi	2491 ± 13		SHRIMP	Nutman et al. (2011)
XHG06	Amphibolite	Xihangou	2543 ± 9	2503 ± 13	CAMECA	Lü et al. (2012)
WZZ06	Meta-felsic volcanic rock	Xihangou	2511 ± 12		CAMECA	Lü et al. (2012)
11JD02-3	Granitic gneisses	Hanerzhaung	2614 ± 18	2461 ± 51	LA-ICP-MS	Guo et al. (2013)
11JD06-3	Two-pyroxene granulite	Eastern Hebei	2525 ± 7	2428 ± 41	LA-ICP-MS	Guo et al. (2013)
11JD07-11	Garnet mafic granulite	Eastern Hebei	2518 ± 15	2470 ± 15	LA-ICP-MS	Guo et al. (2013)
11JD12-1	Two-pyroxene granulite	Eastern Hebei	2523 ± 6		LA-ICP-MS	Guo et al. (2013)
11JD17-2	Two-pyroxene granulite	Eastern Hebei	2609 ± 26	2546 ± 25	LA-ICP-MS	Guo et al. (2013)
QX03	Gabbroic gneiss	Qingyangshu	2536 ± 2		EVAP	Geng et al. (2006)
ZH10	Trondhjemitic gneiss	Qiuhuayu	2527 ± 2	2436	SHRIMP	Geng et al. (2006)
QX05	Tonalitic gneiss	Xiaoguanzhuang	2495 ± 8	2469 ± 33	SHRIMP	Geng et al. (2006)
TP19	Tonalitic gneiss	Longwan	2524 ± 12	2418 ± 31	SHRIMP	Geng et al. (2006)
TP22	Charnockite	Yujuzhai	2505 ± 2	2439 ± 5	EVAP	Geng et al. (2006)
CZ02	Tonalitic gneiss	Cuizhangzi	2492 ± 5	2425 ± 9	EVAP	Geng et al. (2006)
LL07	Granodioritic gneiss	Lucao (Lulong)	2551 ± 6		EVAP	Geng et al. (2006)
NC2022-1/2	Meta-gabbro	Cuizhangzi	2505 ± 2		SGD	Kusky et al. (2001)
Unknown	Tonalitic gneiss	Eastern Hebei	2576 + 29/− 23		CMM	Wu et al. (1991)
Unknown	Granitic gneiss	Eastern Hebei	2592 ± 10		CMM	Wu et al. (1991)
Unknown	Biotite paragneiss	Eastern Hebei		2512 + 57/47	CMM	Wu et al. (1991)
Unknown	Meta-silstone	Eastern Hebei		2497 ± 2	SGD	Wu et al. (1991)
Unknown	Garnet felsic gneiss	Eastern Hebei		2514 + 23/-19	CMM	Wu et al. (1991)
Unknown	Anatectic charnockite	Eastern Hebei	2538 + 41/−36		CMM	Wu et al. (1991)
Unknown	Anatectic charnockite	Eastern Hebei	2526 + 30/−26		CMM	Wu et al. (1991)

(Continued)

Table 2.2 (Continued)

Sample No.	Rock Type	Location	Igneous and Protolithic Age (Ma)	Metamorphic Age (Ma)	Method	Sources
Unknown	Anatectic charnockite	Eastern Hebei	$2513 + 9/-7$		CMM	Wu et al. (1991)
Unknown	Anatectic granite	Eastern Hebei	2484 ± 11	2484 ± 11	SGD	Wu et al. (1991)
Unknown	Anatectic granite	Eastern Hebei	2495 ± 1	2495 ± 1	CMM	Wu et al. (1991)
Unknown	Mafic dyke	Eastern Hebei	$2546 + 29/-24$		CMM	Wu et al. (1991)
CF84-48	Granodiorite	Naoyumen	$2454 + 29/-25$		CMM	Liu et al. (1990)
CF85-73	Gabbroic diorite	Mangshan	$2499 + 8/-7$ 2498 ± 3		CMM SGD	Liu et al. (1990) Liu et al. (1990)
CF85-71	Granodiorite	Mangshan	2494 ± 2 2484 ± 11		CMM SGD	Liu et al. (1990) Liu et al. (1990)
CF84-28A/ 44	Pink granites	Caozhuang	2596 ± 9		CMM	Liu et al. (1990)
CF85-62	Monzodiorite	Naoyumen	2495 ± 4		CMM	Liu et al. (1990)
CF87-9	Charnockite	Yangyashan	$2513 + 9/-7$		CMM	Liu et al. (1990)
Unknown	Tonalitic gneiss	Eastern Hebei	2502 ± 12		CMM	Sun (1984)
Unknown	Tonalitic gneiss	Eastern Hebei	2480 ± 20		CMM	Pidgeon (1980)
Jianping (Western Liaoning)						
OCY10-1	Tonalitic gneiss	Jianping	2511 ± 7	2475 ± 15	LA-ICP-MS	Wang et al. (2012a)
OYX01-4	Granodioritic gneiss	Jianping	2521 ± 9		LA-ICP-MS	Wang et al. (2012a)
OFX11-2	Trondhjemitic gneiss	Jianping	2517 ± 13	2442 ± 11	LA-ICP-MS	Wang et al. (2012a)
YX05-1	Granodioritic gneiss	Jianping	2494 ± 18		LA-ICP-MS	Wang et al. (2012a)
OCY31-2	Tonalitic gneiss	Jianping	2516 ± 6	2465 ± 13	LA-ICP-MS	Wang et al. (2013b)

OCY33-1	Granodioritic gneiss	Jianping	2519 ± 11	2492 ± 19	LA-ICP-MS	Wang et al. (2013b)
CY31-2	Tonalitic gneiss	Jianping	2513 ± 10		LA-ICP-MS	Wang et al. (2013b)
CY46-1	Tonalitic gneiss	Jianping	2506 ± 12		LA-ICP-MS	Wang et al. (2013b)
OCY37-1	Granodioritic gneiss	Jianping	2527 ± 17	2478 ± 23	LA-ICP-MS	Wang et al. (2013b)
OCY46-1	Tonalitic gneiss	Jianping	2532 ± 7	2468 ± 19	LA-ICP-MS	Wang et al. (2013b)
CY36-1	Monzogranitic gneiss	Jianping	2496 ± 12	2457 ± 22	LA-ICP-MS	Wang et al. (2013b)
CY026	Dioritic gneiss	West Jianping	2512 ± 15	2478 ± 9	LA-ICP-MS	Liu et al. (2011a)
CY031-2	Gabbroic gneiss	Northeast Jianping	2513 ± 10	2476 ± 13	LA-ICP-MS	Liu et al. (2011a)
CY032-3	Tonalitic gneiss	Xiguanyingzi	2555 ± 7	2512 ± 12	LA-ICP-MS	Liu et al. (2011a)
CY037-2	Enderbitic gneiss (TTG)	Dayingzi	2538 ± 29		LA-ICP-MS	Liu et al. (2011a)
CY046-1	Quartz dioritic gneiss	West Jianping	2495 ± 9	2457 ± 11	LA-ICP-MS	Liu et al. (2011a)
CY054-1	Enderbitic gneiss	Jianping Town	2510 ± 2	2469 ± 6	LA-ICP-MS	Liu et al. (2011a)
CY055-3	Felsic paragneiss (?)	Jianping Town	2550 ± 4	2471 ± 7	LA-ICP-MS	Liu et al. (2011a)
Ji9	Tonalitic gneiss	Tabei Gold Mine	2495.0 ± 0.4		EVAP	Kröner et al. (1998)
Ji37	Tonalitic gneiss	Zhaojiadian	2473.8 ± 0.6		EVAP	Kröner et al. (1998)
Ji17	Trondhjemitic gneiss	Shaoguozhangzi	2503.9 ± 0.6		EVAP	Kröner et al. (1998)
Ji1	Tonalitic gneiss	Antaigou	2502.7 ± 0.4		EVAP	Kröner et al. (1998)
Ji2	Charnockite	Antaigou	2500.1 ± 0.3		EVAP	Kröner et al. (1998)
Ji11	Meta-dacite	Shaoguozhangzi	2504.5 ± 0.5		EVAP	Kröner et al. (1998)
Ji12	Charnockitic gneiss	Shaoguozhangzi	2501.5 ± 0.5 2501 ± 12	2483.5 ± 0.5	EVAP SHRIMP	Kröner et al. (1998) Kröner et al. (1998)
Ji20	Tonalitic gneiss	Shaoguozhangzi	2521.3 ± 0.4		EVAP	Kröner et al. (1998)
Ji238	Tonalitic gneiss	Zhaojiadian	2503.3 ± 0.5		EVAP	Kröner et al. (1998)
6-34-4	Olivine pyroxenite	NE Jianping		2489.6 ± 0.5	EVAP	Kröner et al. (1998)

(Continued)

Table 2.2 (Continued)

Sample No.	Rock Type	Location	Igneous and Protolithic Age (Ma)	Metamorphic Age (Ma)	Method	Sources
Ji6	Magnetite quartzite	Antaigou		2487 ± 1	SHRIMP	Kröner et al. (1998)
Ji7	Magnetite quartzite	Antaigou		2487 ± 2	SHRIMP	Kröner et al. (1998)
Ji33	Meta-greywacke	Shangshijindaba		2485.5 ± 0.5	EVAP	Kröner et al. (1998)
6-C-4	Pink granite	Changgao Gold Mine	2472.5 ± 0.5	2472.5 ± 0.5	EVAP	Kröner et al. (1998)
Fuxin–Suizhong (Western Liaoning)						
0518-1	Granitic gneiss	Taili (Suizhong)	2522 ± 21		LA-ICP-MS	Zheng et al. (2009a, b)
YX011-2	Andesitic (TTG?) gneiss	Fuxin	2567 ± 27	2493 ± 7	LA-ICP-MS	Wang et al. (2011b)
FX009-1	Andesitic (TTG?) gneiss	Fuxin	2522 ± 8	2493 ± 2	LA-ICP-MS	Wang et al. (2011b)
FX013-2	Andesitic (TTG?) gneiss	Fuxin	2589 ± 16	2502 ± 13	LA-ICP-MS	Wang et al. (2011b)
Southern Liaoning (within the Jiao–Liao–Ji Belt)						
DD09-2	Na-rich gneiss (TTG)	Dachangshan Island	2541 ± 9		LA-ICP-MS	Meng et al. (2013a)
DD09-3	Na-rich gneiss (TTG)	Dachangshan Island	2537 ± 16		LA-ICP-MS	Meng et al. (2013a)
DD09-6	Na-rich gneiss (TTG)	Dachangshan Island	2544 ± 14		LA-ICP-MS	Meng et al. (2013a)
DD09-8	Na-rich gneiss (TTG)	Dachangshan Island	2541 ± 10		LA-ICP-MS	Meng et al. (2013a)
DD29-1	Na-rich gneiss (TTG)	Guanglu Island	2544 ± 9		LA-ICP-MS	Meng et al. (2013a)
DD13-1	Kf-granite	Changshan Island	2516 ± 11		LA-ICP-MS	Meng et al. (2013a)
DD14-3	Kf-granite	Changshan Island	2514 ± 15		LA-ICP-MS	Meng et al. (2013a)
DD14-4	Kf-granite	Changshan Island	2517 ± 17		LA-ICP-MS	Meng et al. (2013a)
DD10-1	Garnet–muscovite schist	Dachangshan Island	2549 ± 14		LA-ICP-MS	Meng et al. (2013b)
DD20-1	Garnet–muscovite schist	Zhangzidao Island	2548 ± 17		LA-ICP-MS	Meng et al. (2013b)

Sample	Lithology	Location	Age	Age	Method	Reference
HY01-1	Garnet–muscovite schist	Haiyang Island	2541 ± 50		LA-ICP-MS	Meng et al. (2013b)
HY03-1	Garnet–muscovite schist	Haiyang Island	2506–2242		LA-ICP-MS	Meng et al. (2013b)
Anshan-Benxi						
WTS1	Amphibolite	Benxi	2523 ± 12	2483 ± 129	LA-ICP-MS	Cui et al. (2013)
A0420	Felsic gneiss (TTG?)	Gongchangling (Anshan)	~2530		SHRIMP	Wan et al. (2013)
A0531	Gneissic monzogranite	Xiangshangou (Anshan)	~2520		SHRIMP	Wan et al. (2013)
A0713	Syenogranite	Qidashan (Anshan)	~2500		SHRIMP	Wan et al. (2013)
WTS	Amphibolite	Waitoushan	2530 ± 121		Cameca	Dai et al. (2012)
Unknown	Tonalitic gneiss	Anshan	2518 ± 12		SHRIMP	Song et al. (1996)
Unknown	Anatectic granite	Eastern Hebei	2474 ± 13	2474 ± 13	SHRIMP	Wu et al. (1991)
Unknown	Paragneiss	Anshan	2518 + 44/−39		CMM	Wu et al. (1991)
Northern Liaoning						
MG-48	Trondhjemiteic gneiss	Chaihe	2553 ± 7	2515 ± 23	SHRIMP	Grant et al. (2009)
MG-146	Trondhjemiteic gneiss	Chaihe	2559 ± 7	2497 ± 7	SHRIMP	Grant et al. (2009)
MG-141	Tonalitic gneiss	Xianjinchang	2534 ± 4	2510 ± 7	SHRIMP	Grant et al. (2009)
MG-47	Syenogranite	Hongshilazi	2502 ± 11	<2493	SHRIMP	Grant et al. (2009)
LQ0107	Fine-grained gneiss	Xiaolaihe Iron Mine	2515 ± 6		SHRIMP	Wan et al. (2005b)
LF0107	Fine-grained gneiss	Tangtu (Fushun)	2510 ± 7		SHRIMP	Wan et al. (2005b)
LQ0104	Fine-grained gneiss	Qingyuan		2479 ± 5	SHRIMP	Wan et al. (2005b)
LF0106	TTG gneiss	Shangma (Fushun)	2528 ± 227	2477 ± 13	SHRIMP	Wan et al. (2005b)
LQ0110	Gneissic granite	East Xiaolaihe	2556 ± 18	2469 ± 139	SHRIMP	Wan et al. (2005b)
Unknown	Tonalitic gneiss	Shangfeidi (Kaiyuan)	2522 ± 15		SHRIMP	Chen et al. (2006)
Unknown	Granitoid gneiss	Hongtoushan (Qingyuan)	2520 ± 16		SHRIMP	Li and Shen (2000)
Unknown	Granitoid gneiss	Paozigou (Qingyuan)	2519 ± 77		SHRIMP	Li and Shen (2000)

(Continued)

Table 2.2 (Continued)

Sample No.	Rock Type	Location	Igneous and Protolithic Age (Ma)	Metamorphic Age (Ma)	Method	Sources
Unknown	Tonalitic gneiss	Qingyuan	2511 ± 1		SGD	Peucut et al. (1986)
Southern Jilin						
Unknown	Amphibolite	Jiapigou	2525 ± 12	2501 ± 13	SGD	Dai et al. (1990)
Unknown	Granitoid gneiss	Baishan	2573 ± 43		SGD	Shen et al. (1998)
Unknown	Syntectonic Kf-granite	Jiapigou	2505 ± 14		SGD	Shen et al. (1998)
Unknown	Anatectic charnockite	Jiapigou	2491 ± 26	2491 ± 26	SGD	Shen et al. (1998)
Unknown	Anatectic quartz dyke	Jiapigou	2475 ± 19	2475 ± 19	SGD	Shen et al. (1998)
Unknown	Anatectic quartz dyke	Jiapigou	2469 ± 33	2469 ± 33	SGD	Shen et al. (1998)
Unknown	Tonalitic gneiss	Jiapigou	2521 ± 14		SGD	Li and Shen (2000)
Unknown	Pink granite	Jiapigou	2497 ± 27		SGD	Li and Shen (2000)
Yishui						
10SD07-2	Metapelitie	Beixiazhuang (Yishui)	2543 ± 6.3	2507 ± 9.4	LA-ICP-MS	Wu et al. (2013a)
10SD07	Charnockite	Mashan pluton (Yishui)	2528 ± 8.3	2508 ± 9.3	LA-ICP-MS	Wu et al. (2013a)
11SD01-2	Garnet charnockite	Caiyu pluton (Yishui)	2559 ± 16	2523 ± 17	LA-ICP-MS	Wu et al. (2013a)
11SD02-16	Charnockite	Xueshan pluton (Yishui)	2535 ± 11	2509 ± 17	LA-ICP-MS	Wu et al. (2013a)
10SD09	Granodioritic gneiss	Mashan pluton (Yishui)	2558.7 ± 8.6	2508 ± 12	LA-ICP-MS	Wu et al. (2013a)
10SD06-4	Granodioritic gneiss	Caiyu pluton (Yishui)	2570 ± 27	2499 ± 7.9	LA-ICP-MS	Wu et al. (2013a)
11SD02-18	Anatectic charnockite	Xueshan pluton (Yishui)		2503 ± 9	LA-ICP-MS	Wu et al. (2013a)
08YS-7	Monzodiorite	Yishui	2543 ± 8		LA-ICP-MS	Peng et al. (2012b)
08YS-43	Monzodiorite	Yishui	2544 ± 4		LA-ICP-MS	Peng et al. (2012b)

Unknown	Enderbite	Tangjiahe (Yishui)	2582 + 25/19		CMM	Su et al. (1999)
YS95-58	Tonalitic gneiss	Caiyu pluton	2562 ± 14	2518 ± 153	SHRIMP	Shen et al. (2004)
YS99-16	Monzogranite	Dashan pluton	2545 ± 10	2508 ± 15	SHRIMP	Shen et al. (2004)
YS9566	Granodioritic gneiss	Mashan pluton (Yishui)	2538 ± 6		SHRIMP	Shen et al. (2007)
YS9573	Trondhjemitic gneiss	Xueshan pluton (Yishui)	2532 ± 6		SHRIMP	Shen et al. (2007)
YS0901-1	Mafic granulite	Qinglongyu (Yishui)	2552–2550	2498.4 ± 7.6	SHRIMP	Zhao et al. (2013)
YS0902-2	Mafic granulite	Qinglongyu (Yishui)	2560 ± 6.6		SHRIMP	Zhao et al. (2013)
YS06-19	Amphibolite	Yishui		2522 ± 5	SHRIMP	Zhao et al. (2009a)
YS06-41	Amphibolite	Yishui		2497 ± 4	SHRIMP	Zhao et al. (2009a)
YS06-40	Mafic granulite	Yishui		2514 ± 5	SHRIMP	Zhao et al. (2009a)
YS06-45	Mafic granulite	Yishui		2485 ± 10	SHRIMP	Zhao et al. (2009a)
YS06-49	Mafic granulite	Yishui		2493 ± 12	SHRIMP	Zhao et al. (2009a)
YS06-29	Amphibolite	Yishui	2531 ± 16		SHRIMP	Zhao et al. (2009a)
YS06-30	Granitic gneiss	Lingshan pluton (Yishui)	2530 ± 7		SHRIMP	Zhao et al. (2008c)
YS06-48	Granitic gneiss	Lingshan pluton (Yishui)	2531 ± 8		SHRIMP	Zhao et al. (2008c)
YS06-29	Amphibolite	Lingshan pluton (Yishui)	2531 ± 5		SHRIMP	Zhao et al. (2008c)
QZK1201-1	Amphibolite	Yangzhuang	2615 ± 61		LA-ICP-MS	Lai and Yang (2012)
QZK1201-1	Biotite quartz schist	Yangzhuang	<2527 ± 66		LA-ICP-MS	Lai and Yang (2012)
QZK401-4	Migmatite	Yangzhuang	2469 ± 34	2469 ± 34	LA-ICP-MS	Lai and Yang (2012)
Luxi (Western Shandong)						
SD0510	Monzogranite	Yuhuangding (Taishan)	2561 ± 23		SHRIMP	Lu et al. (2008a)
SD0504	Tonalite	Huangqian reservoir	2557 ± 20		SHRIMP	Lu et al. (2008a)
SD0611	Quartz diorite	Dazhongqiao (Taishan)	2551 ± 28		LA-ICP-MS	Lu et al. (2008a)
SD0501	Granodiorite	Xiaojinkou (Taishan)	2560 ± 14		SHRIMP	Lu et al. (2008a)

(Continued)

Table 2.2 (Continued)

Sample No.	Rock Type	Location	Igneous and Protolithic Age (Ma)	Metamorphic Age (Ma)	Method	Sources
SD0509	Monzogranite	Aolaishan	2520 ± 8.5		SHRIMP	Lu et al. (2008a)
SD0514	Granodiorite	Yuke	2525 ± 33		SHRIMP	Lu et al. (2008a)
SD0606	Gneissic quartz diorite	Yuhuangding (Taishan)	2523 ± 16		LA-ICP-MS	Lu et al. (2008a)
SD0609	Monzodiorite	Zhongtianmen (Taishan)	2479 ± 65		LA-ICP-MS	Lu et al. (2008a)
SD0612	Fine-grained diorite	Puzhaosi	2481 ± 17		SHRIMP	Lu et al. (2008a)
JN0740	Amphibolite	Taishan	2570 ± 18		LA-ICP-MS	Jiang et al. (2010)
Unknown	Granodiorite	Guimengding (Mengshan)	2539 ± 17		SHRIMP	Wang et al. (2008)
08YS-197	Granodiorite	Wop (Taishan)	2548 ± 16		LA-ICP-MS	Peng et al. (2012b)
08YS-146	Monzodiorite	Jiangjundi (Zoucheng)	2547 ± 8		LA-ICP-MS	Peng et al. (2012b)
08YS-112	Monzogranite	Wanghailou (Feixian)	2543 ± 16		LA-ICP-MS	Peng et al. (2013a)
08YS-105	Monzogranite	Hujiazhuang (Feixian)	2517 ± 21		LA-ICP-MS	Peng et al. (2013a)
08YS-142	Monzogranite	Xiajiazhuan (Feixian)	2462 ± 18		LA-ICP-MS	Peng et al. (2013a)
08YS-98	Amphibolite	Hujiazuang Village	2532 ± 15		LA-ICP-MS	Peng et al. (2013b)
S0723	Hb–Bi gneiss (volcanic)	Hanwang (Shancaoyu F.)	2534 ± 9		SHRIMP	Wan et al. (2012a)
S0737	Bi gneiss (volcanic)	Qixingtai (Shancaoyu F.)	2553 ± 9		SHRIMP	Wan et al. (2012a)
TS0924-1	Bi gneiss (volcanic)	Qixingtai (Shancaoyu F.)	2551 ± 10		SHRIMP	Wan et al. (2012a)
TS0928	Bi gneiss (volcanic)	Qixingtai (Shancaoyu F.)	2551 ± 7		SHRIMP	Wan et al. (2012a)
S0736	Hb gneiss (volcanic)	Qixingtai (Yanglingguan)	2534 ± 9		SHRIMP	Wan et al. (2012a)
S0307	Garnet-biotite gneiss	Shancaoyu	2611 ± 7		SHRIMP	Wan et al. (2012a)
SY0319	Metavolcanic pebble	Upper Liuhang F.	2553 ± 10		SHRIMP	Wan et al. (2012a)

Sample	Description	Location	Age	Age	Method	Reference
SY0320	Matrix in conglomerate	Upper Liuhang F.	2587 ± 16		SHRIMP	Wan et al. (2012a)
TS09104	Felsic metavolcanics	Upper Liuhang F.	2524 ± 7		SHRIMP	Wan et al. (2012a)
S0307	Biotite gneiss	Shancaoyu F.	2572 ± 16		SHRIMP	Wan et al. (2012a)
TS09105	Biotite gneiss	Shancaoyu F.	2544 ± 6		SHRIMP	Wan et al. (2012a)
S0720	Hornblende gneiss	Shancaoyu F.	2529 ± 6		SHRIMP	Wan et al. (2012a)
S0708	Biotite gneiss	Shancaoyu F.	2572 ± 16		SHRIMP	Wan et al. (2012a)
S0719	Biotite gneiss	Mengliangpu	2553 ± 18		SHRIMP	Wan et al. (2012a)
S0712	Biotite gneiss	Mengliangpu	2548 ± 9		SHRIMP	Wan et al. (2012a)
S0844-2	Meta-felsic volcanics	Jining (Jining Group)	2561 ± 15		SHRIMP	Wang et al. (2010b)
YN3	Dioritic gneiss	Mengshan	2536 ± 26		SHRIMP	Hou et al. (2008a)
06TK-1	High-mg dioritic gneiss	Taishan	2540 ± 9		LA-ICP-MS	Wang et al. (2009b)
Qixia (Within the Paleoproterozoic Jiao–Liao–Ji Belt)						
10SD10-2	Garnet amphibolite	3 km NW Qixia City	2546 ± 12	2458 ± 9 / 1887 ± 55	LA-ICP-MS	Wu et al. (2013b)
10SD19-7	Garnet amphibolite	18 km SE Qixia City	2559 ± 10	2522 ± 11 / 1929–1836	LA-ICP-MS	Wu et al. (2013b)
10SD27-9	Garnet amphibolite	5 km SE Qixia City		2473 ± 20 / 1838 ± 25	LA-ICP-MS	Wu et al. (2013b)
10SD13-2	Amphibolite	12 km SE Qixia City	2588 ± 16 / 2571 ± 18	2469 ± 30 / 1865 ± 13	LA-ICP-MS	Wu et al. (2013b)
10SD16-1	Trondhjemitic gneiss	Daliujia Village	2526 ± 23	2501 ± 16 / 1849 ± 12	LA-ICP-MS	Wu et al. (2013b)
10SD19-2	Amphibolite	18 km SE Qixia City	2555 ± 11	2506–2459 / 1948 ± 30	LA-ICP-MS	Wu et al. (2013c)
10SD26-1	Granodioritic gneiss	8 km SE Qixia City	2573 ± 12	2493 ± 14	LA-ICP-MS	Wu et al. (2013c)
QX13-3	Tonalitic gneiss	Xiliu Town	2548 ± 12	2504 ± 16	LA-ICP-MS	Liu et al. (2013b)

(Continued)

Table 2.2 (Continued)

Sample No.	Rock Type	Location	Igneous and Protolithic Age (Ma)	Metamorphic Age (Ma)	Method	Sources
				1863 ± 41		
QX69-1	Granitoid gneiss	Yangjiaqiao Village	2522 ± 55	2510 ± 54 1861−1857	LA-ICP-MS	Liu et al. (2013b)
QX27-2	Granitoid gneiss	Sunjiayanhou	2544 ± 15		LA-ICP-MS	Liu et al. (2013b)
QX66-1	Tonalitic gneiss	Quanli Village	2564 ± 12		LA-ICP-MS	Liu et al. (2013b)
Xenoliths or Zircons from Phanerozoic Intrusions						
08JG18	Garnet-bearing gneiss	Jiagou intrusion	2552 ± 13	2475 ± 15	SHRIMP	Wang et al. (2012b)
SL1	Metamorphic zircons	Mengyin Kimberlite		2524 ± 5	LA-ICP-MS	Zheng et al. (2009a)
HQ5	Metamorphic zircons	Mengyin Kimberlite		2540 ± 30	LA-ICP-MS	Zheng et al. (2009a)

SHRIMP, Sensitive high resolution ion microprobe; CAMECA, Cameca ion microprobe; EVAP, single grain evaporation; SGD, single grain dissolution; CMM, conventional multigrain method; EPMA, electron probe microanalysis; LA-ICP MS, Laser Ablation Inductively Coupled Plasma Mass Spectrometry.

rocks, ~2.5 Ga syntectonic charnockites (granulite-facies terrains) and granites (amphibolite-facies terrains), with minor volumes of 2.55–2.50 Ga bimodal volcanic rocks and sedimentary rocks (Pidgeon, 1980; Jahn and Zhang, 1984; Zhai et al., 1985, 1990, 2003; Jahn et al., 1988; Jahn, 1989; Jahn and Ernst, 1990; Shen and Qian, 1992; Shen et al., 1992; Kröner et al., 1998; Zhao et al., 1998; Geng et al., 2006, 2012; Wan et al., 2012b; Zhang et al., 2011,2012a). The major geological features of the 2.55–2.50 Ga rock associations in the Eastern Block can be summarized as follows:

1. Spatially, the 2.55–2.50 Ga rock associations in the Eastern Block are distributed over the whole block but do not show any systematic age progression across the ~800 km wide (Figure 2.1). These 2.55–2.50 Ga rock associations are exposed as high- to medium-grade gneissic complexes in the Miyun, Eastern Hebei, Jianping, Yishui, Qixia, Fuxin-Suizhong, North Liaoning, and Southern Jilin areas and as low- to medium-grade granite-greenstone terranes in the Western Shandong (Luxi), Southern Liaoning, and Anshan-Benxi areas (Figure 2.1). As given in Table 2.2, the granitoid (TTG) gneisses in the high-grade complexes and low-grade granite-greenstone terranes were formed nearly coevally, all at 2.55–2.50 Ga. The occurrence of temporally and petrologically similar rocks across a ~800 km wide area in the Eastern Block of the NCC is impressive, which is inconsistent with what we have observed in Phanerozoic accretionary terranes.

2. Lithologically, most of the 2.6–2.5 Ga greenstones in the Eastern Block consist of bimodal volcanic assemblages, in which basalt and ultramafic rocks are associated with dacite, rhyodacite, and rhyolite. Some of the ultramafic rocks are komatiites or komatiitic rocks, which have been found in the Hongtoushan greenstone belt in Northern Liaoning (Chen, 1983), the Zunhua–Qianxi greenstone belt in Eastern Hebei (Zhang et al., 1980; Li, 1982), and Longgang greenstone belt in Southern Jilin (Liu, 2001), though the most typical komatiites with spinifex textures in the Luxi greenstone belt are formed at 2.8–2.7 Ga (Cheng and Xu, 1991; Polat et al., 2006; Cheng and Kusky, 2007; Wan et al., 2011a).

3. All of the 2.55–2.50 Ga rock associations in the Eastern Block underwent regional metamorphism at ~2.5 Ga (Pidgeon, 1980; Jahn and Zhang, 1984; Jahn et al., 1988, 2008; Kröner et al., 1998; Geng et al., 2006, 2010, 2012; Yang et al., 2008; Grant et al., 2009;

Li et al., 2010b; Liu et al., 2011a,b; Nutman et al., 2011; Zhang et al., 2012d). Associated with the 2.5 Ga metamorphic event was the emplacement of syntectonic charnockite-enderbite suites by recrystallization of granitoid (TTG) gneisses (in granulite-facies terrains) and K-rich granites (in amphibolites-facies terrains; Wan et al., 2012b). In the most high-grade complexes in the Eastern Block, the 2.55–2.50 Ga rocks preserve both magmatic and metamorphic zircons, of which the latter occur as either overgrowth rims surrounding older magmatic zircon cores or single grains, and are structureless (Figure 2.7), highly luminescent and very low in Th and U contents. These features make them distinctly different from the magmatic zircons that are generally characterized by oscillatory zoning (Figure 2.7), low luminescence, and comparatively high Th and U contents. The metamorphic zircons from both the 2.55–2.50 Ga high-grade TTG gneisses and mafic granulites/amphibolites yield similar $^{207}Pb/^{206}Pb$ ages in the range from 2520 to 2450 Ma (Table 2.2), which clearly demonstrate that the 2.55–2.50 Ga rock associations in the Eastern Block underwent regional metamorphism at ~2.5 Ga.

4. The 2.55–2.50 Ga granitoid (TTG) gneisses in the Eastern Block are characterized by near-contemporaneity of granitoid intrusive and metamorphic events, with the peak of metamorphism reaching shortly after the widespread intrusion of granitoid suites (Jahn and Zhang, 1984; Kröner et al., 1998; Zhao et al., 1998, 2001a; Geng et al., 2006, 2010, 2012; Yang et al., 2008; Grant et al., 2009; Wu et al., 2013a,b,c). Jahn and Zhang (1984) are the first to recognize

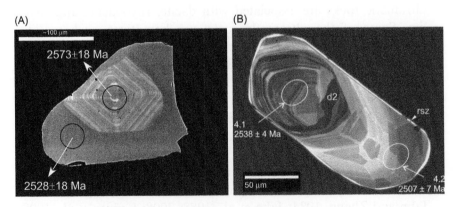

Figure 2.7 CL images of igneous zircon cores and metamorphic rims from (A) granulites (Zheng et al., 2004b) and (B) high-grade TTG gneisses (Grant et al., 2009) in the Eastern Block of the NCC, which show near-contemporaneity of initial magmatism and high-grade regional metamorphism.

the near-contemporaneity of TTG-intrusive and granulite-facies metamorphic events in Eastern Hebei. On the basis of their own Rb–Sr and Sm–Nd isotopic results and previous U–Pb zircon ages (Pidgeon, 1980), Jahn and Zhang (1984) concluded that the emplacement of high-grade TTG gneisses (felsic granulites) in Eastern Hebei was followed shortly, within less than 100 Ma, by a granulite-facies metamorphic event at ~2.5 Ga. This conclusion was further supported by later isotopic data from this area (Liu et al., 1990; Wu et al., 1991; Geng et al., 2006; Yang et al., 2008; Nutman et al., 2011; Lü et al., 2012; Guo et al., 2013). Such near-contemporaneity of TTG-intrusive and granulite-facies metamorphic events was also revealed in other Neoarchean gneissic complexes in the Eastern Block of the NCC (Kröner et al., 1998; Shen et al., 1998, 2004; Zhao et al., 1998, 2009a, 2013; Wan et al., 2005b; Geng et al., 2006; Grant et al., 2009; Liu et al., 2011a, 2013b; Wang et al., 2011b, 2012a, 2013b; Cui et al., 2013; Meng et al., 2013a; Wu et al., 2013a,b,c). In the Jianping Complex, for example, SHRIMP and LA-ICP-MS zircon dates obtained by Kröner et al. (1998), Liu et al. (2011a), and Wang et al. (2013b) document an evolutionary history that began with deposition of a cratonic supracrustal sequence some 2551–2522 Ma ago and widespread intrusion of granitoid rocks, beginning at 2550 Ma and reaching a peak at about 2500 Ma, shortly followed by regional granulite-facies metamorphism, transforming the existing rocks into granulites, charnockites, and enderbites between 2492 and 2457 Ma (Table 2.2). This means that the granitoid (TTG) intrusives in the Jianping Complex are formed less than about 50 Ma prior to their deformation and granulite-facies metamorphism (Kröner et al., 1998; Liu et al., 2011a; Wang et al., 2013b). In the Northern Liaoning Complex, the TTG suites were emplaced in the period 2559–2553 Ma and encountered granulite-facies metamorphism at 2517–2490 Ma (Table 2.2; Wan et al., 2005b; Grant et al., 2009; Figure 2.7A). In the Southern Liaoning Complex, Zheng et al. (2004b) revealed that the time gap between the magmatic and metamorphic zircons from Neoarchean granulite xenoliths enclosed in the Paleozoic kimberlites is less than 20 Ma (Figure 2.7B). This is the same case with the Luxi, Qixia, and Southern Jilin complexes where the regional metamorphism occurred at ~2.5 Ga, shortly after the emplacement of granitoid (TTG) suites in the period 2.55–2.50 Ga (Table 2.2; Shen et al., 1998, 2004; Li and Shen,

2000; Lu et al., 2008b; Jahn et al., 2008; Zhao et al., 2009a, 2013; Wan et al., 2011a,b; Liu et al., 2013a,b; Wu et al., 2013a,b,c). Such near-contemporaneity of granitoid intrusive and metamorphic events is unique to the 2.55−2.50 Ga high-grade granitoid (TTG) gneisses in the Eastern Block of the NCC, but it is not the case with the 2.55−2.50 Ga high-grade granitoid (TTG) gneisses in the Trans-North China Orogen where the time gap between the granitoid magmatism and regional metamorphism is about 700 Ma (see details in Chapter 4).

5. The metamorphic evolution of the 2.55−2.50 Ga rock associations in the Eastern Block is exclusively characterized by anticlockwise $P−T$ paths involving isobaric cooling (IBC), reflecting underplating and intrusion of mantle-derived magmas (Zhao et al., 1998, 1999a,b and references wherein). The 2.55−2.50 Ga high-grade TTG gneisses in the Eastern Block commonly contain boudins or sheets of medium-pressure mafic granulites and pelitic gneisses that preserve prograde, peak, and postpeak mineral assemblages. In the mafic granulites, the prograde assemblage is hornblende + plagioclase ± quartz ± biotite, preserved as mineral inclusions within the peak-stage minerals (Figure 2.8A; Zhao, 2001, 2009); the peak assemblage is garnet + orthopyroxene + clinopyroxene + plagioclase, a typical medium-pressure granulite-facies assemblage, whereas the postpeak assemblage is characterized by garnet + quartz symplectic coronas surrounding the pyroxene and plagioclase grains (Figure 2.8B; Zhao et al., 1999a, 2001a) or growing around a garnet grain that formed at the peak stage (Figure 2.8C; Zhao, 2009). In the pelitic gneisses, the prograde assemblage is biotite + plagioclase + quartz ± andalusite; the peak assemblage is garnet + sillimanite + plagioclase + quartz + biotite, whereas the postpeak stage is characterized by the kyanite replacing sillimanite or kyanite cutting the sillimanite lineations (Figure 2.8D; Ge et al., 2003). Since the 1990s, Chinese researchers have used the traditional geothermobarometers to calculate the $P−T$ conditions of the prograde, peak, and postpeak mineral assemblages recognized in the mafic granulites and pelitic gneisses and established their metamorphic $P−T$ paths (Figure 2.9; Cui et al., 1991; He et al., 1991; Liu, 1991; Xu et al., 1992; Li, 1993; Sun et al., 1993a; Chen et al., 1994; Zhao et al., 1998, 1999b, 2001a and references wherein). As shown in Figure 2.9, Neoarchean mafic granulites and pelitic gneisses from different complexes in the Eastern Block, regardless of metamorphic

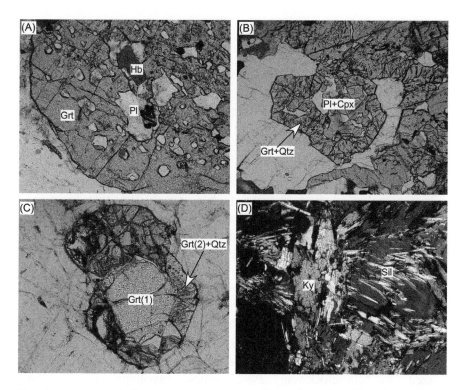

Figure 2.8 Photomicrographs illustrating representative metamorphic reaction textures of the mafic granulites and pelitic gneisses in the Eastern Block (Zhao, 2009). Descriptions see the text. Mineral abbreviations: Cpx, clinopyroxene; Grt, garnet; Ky, kyanite; Pl, plagioclase; Sil, sillimanite; Qtz, quartz.

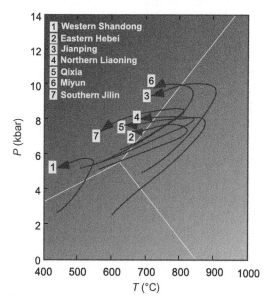

Figure 2.9 P−T paths of Neoarchean metamorphic complexes in the Eastern Block. 1, Western Shandong (Xu et al., 1992; Zhao et al., 1998); 2, Eastern Hebei (He et al., 1991; Liu, 1991; Zhao et al., 1998); 3, Jianping (Cui et al., 1991); 4, Northern Liaoning (Sun et al., 1993a); 5, Eastern Shandong (Li, 1993); 6, Miyun (Chen et al., 1994); 7, Southern Jilin (Ge et al., 2003).

grades and compositions, show remarkably similar metamorphic evolution, which is characterized by anticlockwise $P-T$ paths involving nearly IBC. Although some errors may have existed in the $P-T$ estimates due to the application of inconsistent traditional thermobarometers, the similarity in $P-T$ evolution among different metamorphic complexes in the Eastern Block is clearly not an artifact of thermobarometry, because the inferred $P-T$ paths are not only defined by the $P-T$ estimates, but are also well constrained by metamorphic reaction textures, such as nearly IBC textures represented by the garnet + quartz symplectic coronas surrounding plagioclase and pyroxene in the mafic granulites (Zhao et al., 1999b), and kyanite replacing sillimanite in the pelitic gneisses (Ge et al., 2003). Most recently, Wu et al. (2012, 2013d) applied the THERMOCALC pseudosection modeling technique to reconstruct the metamorphic $P-T$ paths of the Yishui and Northern Liaoning complexes, respectively, and the results show that the metamorphic $P-T$ paths of the two complexes are similar to those reconstructed using the traditional geothermobarometric methods, both characterized by anticlockwise $P-T$ paths involving nearly IBC (Figures 2.10 and 2.11). Generally, the anticlockwise $P-T$ paths involving IBC reflect an origin related to the intrusion and underplating of large amounts of mantle-derived magmas. The intrusion and underplating of mantle magmas not only provide heat for the metamorphism, but also add an adequate volume of mostly mafic material to the crust, leading to both temperature- and pressure increasing during the prograde and peak metamorphic stages. When the underplating of mantle magmas ceases, the crust would not undergo large-scale exhumation or uplift but only experience cooling following the peak metamorphism, forming an anticlockwise $P-T$ path with IBC like those shown in Figure 2.9.

6. The structural patterns of both high-grade gneissic complexes and low- to medium-grade granite-greenstone terranes in the Eastern Block are characterized by domiform structures. In the Eastern Block, the 2.55−2.50 Ga TTG gneiss domes are generally circular, elliptical, or oval, 10−50 km in diameter, and consist of broadly uniform TTG gneisses, locally associated with ∼2.5 Ga syntectonic charnockites or granites in the cores of the domes. The representative domes include the Jinzhou dome in Southern Liaoning (Sun et al., 1992a); the Qianan, Chuizhangzi, and Taipingzhai-

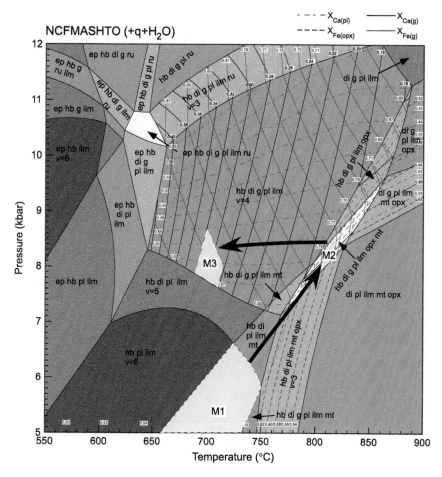

Figure 2.10 Metamorphic P–T path of the Yishui Complex reconstructed using the THERMOCALC pseudosection modeling technique (Wu et al., 2012).

Santunying domes in Eastern Hebei (Figure 2.12A; He et al., 1992); and the Qingyuan and Huadian domes in Northern Liaoning and Southern Jilin (Figure 2.12B; Sun et al., 1993a). Such domiform structures are also dominant in many other Archean cratonic blocks such as Zambabwe, Barberton, Yilgarn, and Pilbara. The origins of the gneiss domes in the Eastern Block are still controversial, with some people believing that they resulted from the superimposition of multiple folds (Bai, 1980), whereas others argue that they were related to the diapiric intrusion of granitoid magmas (Zhao et al., 1998).

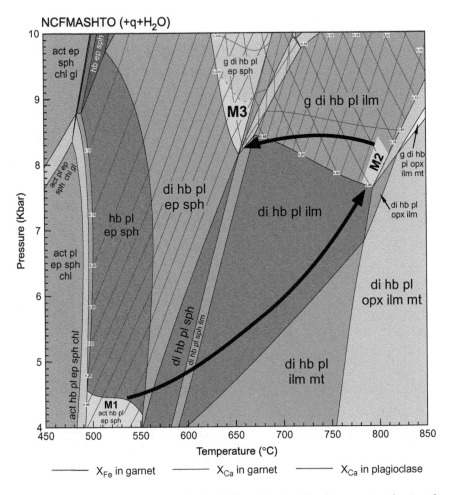

Figure 2.11 Metamorphic P–T path of the Northern Liaoning Complex reconstructed using the THERMOCALC pseudosection modeling technique (Wu et al., 2013d).

2.7 MAJOR ARCHEAN GEOLOGICAL EVENTS AND THEIR NATURE

Table 2.3 lists major geological events that happened in the Eastern Block during Archean times, which are summarized as follows.

2.7.1 Eoarchean Geological Events

The oldest Archean geological event in the Eastern Block was the emplacement of the ~3.8 Ga trondhjemite and diorite into a pre-3.85 Ga (Hadean) crust in the Anshan area, represented by the 3811 ± 4 Ma Dongshan trondhjemitic gneiss, $3804 \pm 5/3802 \pm 11$ Ma

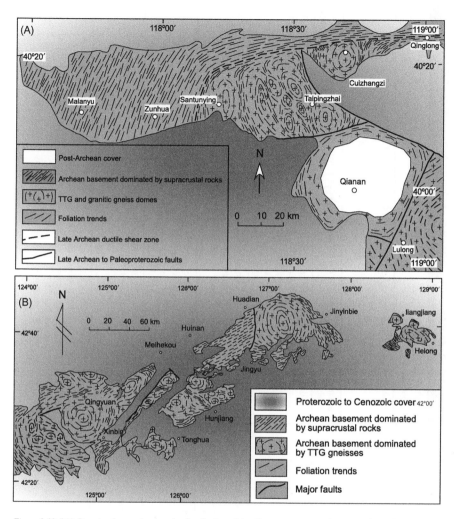

Figure 2.12 (A) Structural map showing the distribution of domiform structures in Eastern Hebei. (B) Structural map showing the distribution of domiform structures in Northern Liaoning and southern Jilin. (A) After He et al. (1992); (B) After Sun et al. (1993a).

Baijiafen trondhjemitic gneiss, 3792 ± 4 Ma Dongshan quartz dioritic gneiss, and 3773 ± 6 Ma Shengousi tronhjemitic gneiss (Liu et al., 1992, 2007, 2008a; Song et al., 1996; Wan et al., 2005a). Most zircons from these Eoarchean rocks have $\varepsilon Hf_{(t)}$ values ranging from -2.07 to $+13.52$ and $T_{DM}(Hf)$ model ages around 3.8–4.0 Ga, with most at ~ 3.9 Ga (Liu et al., 2008a), close to their crystallization ages, indicating that this magmatic event resulted from the partial melting of a juvenile crust. Minor zircons from these rocks have older $T_{DM}(Hf)$

Table 2.3 Major Geological Events in the Eastern Block

Time	Major Geological Events	Spatial Distribution	Geological Records	εHf(t) Value/ T_DM(Hf) Model Age
Hadean (>3.85 Ga)	4.26–4.0 Ga	Formation of a Hadean proto-crust	Anshan	• T_{DM}(Hf) model ages of 4.26–3.85 Ga for zircons from from Eoarchean rocks in Anshan[1,2] • A 4174 ± 48 Ma zircon from the 2523 ± 12 Ma amphibolite in Anshan[3]	
	4.0–3.85 Ga		Eastern Hebei	• 3.86–3.88 Ga detrital zircons from the Caozhuang fuchisite quartzite[4,5]	−0.3 to +4.0/ 3.90–3.94 Ga[4]
Eoarchean (3.85–3.60 Ga)	3.81–3.77 Ga	Emplacement of the ~3.8 Ga trondhjemite in Anshan	Anshan	• 3811 ± 4 Ma Dongshan trondhjemitic gneiss[1] • 3804 ± 5 Ma Baijiafen trondhjemitic gneiss[6] • 3802 ± 11 Ma Baijiafen trondhjemitic gneiss[7] • 3792 ± 4 Ma Dongshan quartz dioritic gneiss[8] • 3773 ± 6 Ma Shenguosi tronhjemitic gneiss[9]	−2.07 to 13.52/ 3.8–4.0 Ga[1]
	3.68–3.62 Ga	1. Formation of supracrustals in Anshan 2. Emplacement of the ~3.6 Ga trondhjemite in Anshan	Anshan	• 3680 ± 19 Ma Dongshan trondhjemites[1] • 3.72–3.62 Ga Baijiafen supracrustals[1] • 3620 ± 23 Ma and 3573 ± 21 Ma Baijiafen trondhjemites[1]	−4.69 to +3.52/ 3.7–4.1 Ga (most around 3.85 Ga)[1]
Eoarchean/ Paleoarchean	~3.55 Ga	Metamorphism in Anshan and Eastern Hebei	Anshan/ Eastern Hebei	• A 3556 ± 6 Ma metamorphic zircon from the Shenguosi tronhjemitic gneiss (Anshan)[1,9] • A 3551 ± 12 Ma metamorphic zircon from the Caozhuang fuchsite quartzite (Eastern Hebei)[6]	
Paleoarchean (3.6–3.2 Ga)	3.57–3.45 Ga	1. Eruption of ~3.5 Ga basalts in Eastern Hebei 2. Emplacement of ~3.45 Ga trondhjemite in Anshan	Eastern Hebei/ Anshan	• 3500 ± 80 Ma, 3470 ± 107 Ma, and 3561 ± 15 Ma Caozhuang amphibolites (Eastern Hebei)[10–12] • 3454 ± 8 Ma and 3448 ± 9 Ma Shenguosi trondhjemitic gneisses (Anshan)[13]	
	3.36–3.3 Ga	1. Emplacement of ~3.3 Ga trondhjemite and granite in Anshan 2. Formation of the Chentaigou supracrustals	Anshan	• 3306 ± 13 Ma Chentaigou granite[14] • 3362–3342 Ma Chentaigou supracrustals[14] • 3342 ± 10 Ma a granitic dyke cutting the Chentaigou supracrustals[14] • 3303 ± 5 Ma Dongshan dioritic gneiss and 3305 ± 7 Ma Dongshan trondhjemitic gneiss[7]	−5.29 to +4.36/ 3.43–3.93 Ga[7] − 0.43 to −4.32/ 3.52–3.66 Ga[7]

Age	Event	Location	Details	$\varepsilon_{Nd}/\varepsilon_{Hf}$ and model ages
Mesoarchean (3.2–2.8 Ga) 3.14–2.86 Ga	1. Emplacement of ~3.0 Ga K-rich granites in Anshan and Eastern Hebei 2. Formation of the Qianan supracrustals	Anshan/ Eastern Hebei	• 3142 ± 7 Ma Lishan granite, 3001 ± 8 Ma West Anshan granite, 2994 ± 8 Ma East Anshan granite, and 2992 ± 10/2983 ± 10/ 2962 ± 4 Ma Tiejiashan K-rich granites (Anshan) [14,15] • >2980 ± 8 Ma Qianan supracrustals[16] • 2980 ± 8 Ma Yangyashan orthogneiss[16]	−7.85 to −2.29/ 3.23–3.48 Ga[15]
2.90–2.85 Ga	3. Emplacement of 2.90–2.85 Ga TTGs	Qixia	• 2909 ± 19 Ma Shilizhuang tonalitic gneiss[22] • 2906 ± 12 Ma Huangyadi grey gneiss[17] • 2892 ± 18 Ma Huangyadi supracrustals[17] • 2865 ± 18 Ma Dachai granodioritic gneiss[18]	+3.2 to +10.2/ 2.9–3.2 Ga[22] 0.8 to +6.4/ 3.39–2.90 Ga[18]
Neoarchean (2.8–2.5 Ga) 2.75–2.65 Ga	A ~2.7 Ga LIP event to form a mafic crust	Whole Eastern Block	• T$_{DM}$(Hf) model ages of 2.8–2.7 Ga for zircons from 2.6–2.5 rocks from the whole Eastern Block[19]	
	Emplacement of TTG and the eruption of volcanic rocks in Luxi and Qixia	Luxi	2.75–2.65 Ga TTG rocks and Taishan Group in the Luxi granite-greenstone terrane[20]	+2.7 to +10.0/ 2.97–2.62 Ga[19]
		Qixia	2.75–2.70 Ga TTG rocks in the Qixia Complex[17,18]	−2.0 to +5.1/ 3.52–2.90 Ga[18]
~2.65 Ga	Metamorphism	Qixia/Luxi	~2.65 Ga metamorphic zircons from the TTG gneisses in Luxi[20]	
2.55–2.50 Ga	Emplacement of TTG and eruption of mafic–felsic volcanics in the whole Eastern Block	Whole Eastern Block	• 2.55–2.50 Ga TTG[21] • 2.6–2.5 Ga mafic–felsic volcanics[21]	−8.8 to +10.1/ 3.2–2.6 Ga (mostly at 2.6–2.8 Ga)[19]
Neoarchean/ Paleoproterozoic (~2.5 Ga) ~2.5 Ga	Regional metamorphism	Whole Eastern Block	• 2.50–2.45 Ga metamorphic ages obtained for Neoarchean rocks in the Eastern Block[21] • ~2.50 anatectic charnockite/pink granite[21]	

Reference sources: (1), Liu et al. (2008a); (2), Wu et al. (2009); (3), Cui et al. (2013); (4), Wu et al. (2005a); (5), Wilde et al. (2008); (6), Liu et al. (1992); (8), Wan et al. (2005a); (9), Liu et al. (2007); (10), Huang et al. (1986); (11), Qiao et al. (1987); (12), Jahn et al. (1987); (13), Wan et al. (2012c); (14), Song et al. (1996); (15), Wan et al. (2007); (16), Wu et al. (1998); (17), Jahn et al. (2008); (18), Wu et al. (2013b); (19), references in Table 2.4; (20), references in Table 2.1; (21), references in Table 2.2; Liu et al. (2013b).

model ages ranging between 4.0 and 4.1 Ga, implying the involvement of a Hadean proto-crust during the partial melting of the juvenile crust to form the trondhjemitic-dioritic magmas. At the late stage of the Eoarchean (3.68−3.62 Ga), some of these Eoarchean rocks were exposed upon the surface where they underwent weathering and erosion, producing sediments that were then deposited in basins to form the sedimentary protoliths of the Baijiafen supracrustals, whose depositional time has been well constrained at 3.68−3.62 Ga by their youngest detrital zircon age of ∼3.68 Ga, and a 3.62 Ga trondhjemitic vein that intrudes the supracrustals (Liu et al., 2008a). Nearly coeval with the deposition of sedimentary protoliths of the Baijiafen supracrustals was emplacement of a ∼3.6 Ga trondhjemites, represented by the 3680 ± 19 Ma Dongshan trondhjemite and the 3620 ± 23−3573 ± 21 Ma Baijiafen trondhjemite. Zircons from these trondhjemites possess $\varepsilon Hf_{(t)}$ values ranging from −4.69 to +3.52 and $T_{DM}(Hf)$ model ages of 3.7−4.1 Ga, with most around 3855 Ma, implying that this trondhjemitic magamtism was derived mainly from the partial melting of the preexisted Hadean−Eoarchean crust with younger juvenile crustal addition at ∼3.65 Ga. At the end of Eoarchean (∼3.55 Ga), the Anshan area experienced a metamorphic event, as evidenced by a 3556 ± 6 Ma metamorphic zircon recognized from the Shengousi tronhjemitic gneiss (Liu et al., 2008a). This end-Eoarchean metamorphic event was also recorded in the Caozhuang fuchsite quartzite of which a 3551 ± 12 Ma metamorphic zircon was recognized (Liu et al., 1992).

2.7.2 Paleoarchean Geological Events

Paleoarchean geological events in the Eastern Block have only been revealed in Anshan and Eastern Hebei (Table 2.3). In Eastern Hebei, the early phase of the Paleoarchean events is marked by the eruption of basalts to form the precursors of the Caozhuang amphibolites which yielded whole-rock Sm−Nd isochron ages of 3500 ± 80 (Huang et al., 1986), 3470 ± 107 (Jahn et al., 1987), and 3561 ± 15 Ma (Qiao et al., 1987). Jahn et al. (1987) showed that the Caozhuang amphibolites are characterized by positive $\varepsilon Nd(T)$ values of +2.7, implying that the protolithic basalt magmas were generated from a mantle source that had undergone long-term depletion, and such a depletion event must have taken place around 4.0 Ga, which may have been related to the extraction of the Hadean continental material as inferred from 3.86−3.88 Ga detrital zircons from the Caozhuang fuchsite quartzite and their $T_{DM}(Hf)$ model ages (Wu et al., 2005a; Wilde et al., 2008; Nutman et al., 2011). In the Anshan area, the early phase of the Paleoarchean events is marked by emplacement of the Shengousi trondhjemites at 3.45 Ga (Table 2.3). The late Paleoarchean events in the Eastern Block only occurred in the Anshan area, marked by emplacement of the Dongshan diorite and trondhjemite and the Chentaigou granite at ∼3.3 Ga (Table 2.3), of which the 3305 ± 7 Ma Dongshan trondhjemite represents the youngest phase (Phase IV) of trondhjemitic magmatism in the area. Zircons from the Chentaigou granite have $\varepsilon Hf_{(t)}$ values ranging from −5.29 to +4.36 and $T_{DM}(Hf)$ model ages of 3.43−3.93 Ga (Table 2.3; Wu et al., 2008), implying that the granite was derived mainly from partial melting of the preexisted Eoarchean crust with juvenile crustal addition. In contrast, all zircons from the Dongshan dioritic and trondhjemitic gneisses possess negative $\varepsilon Hf_{(t)}$ values ranging from −0.43 to −4.32 and $T_{DM}(Hf)$ model ages of 3.52−3.66 Ga (Table 2.3; Wu et al., 2008), implying that these rocks were mainly derived from the recycling of the late Eoarchean to early Paleoarchean crustal rocks. Nearly coeval with the emplacement of the Dongshan trondhjemite-diorite and the Chentaigou granite into the middle to upper crust of the Anshan area was the formation of sedimentary-volcanic rocks that were subsequently transformed into the Chentaigou supracrustals, whose depositional time is restricted to a narrow period between 3362 and 3342 Ma by ages of the youngest detrital zircons and a granitic dyke cutting the supracrustals (Song et al., 1996).

2.7.3 Mesoarchean Geological Events

Mesoarchean geological events in the Eastern Block mainly occurred in the Anshan, Eastern Hebei, and Qixia areas (Figures 2.2−2.4; Table 2.3). An early phase of the Mesoarchean events was emplacement of ∼3.0 Ga granites in Anshan and Eastern Hebei, represented by the ∼3.14 Ga Lishan granite, ∼3.0 Ga West Anshan granite, ∼2.99 Ga East Anshan granite, and 2.99−2.96 Ga Tiejiashan granite in the Anshan area (Song et al., 1996; Wan et al., 2007) and the 2980 ± 8 Ma Yangyashan granite in Eastern Hebei (Wu et al., 1998). Zircons from the Tiejiashan granite have $\varepsilon Hf_{(t)}$ values of −7.85 to −2.29 and $T_{DM}(Hf)$ model ages of 3.23−3.48 Ga (Table 2.3; Wan et al., 2007), implying that this Mesoarchean magmatism most likely resulted from partial melting of late Paleoarchean rocks, whose most possible candidates were the ∼3.3 Dongshan diorite and trondhjemite.

The younger phase (2.9−2.85 Ga) of the Mesoarchean events occurred in the Qixia area of Eastern Shandong, marked by emplacement of the 2906 ± 12 Ma Huangyadi tonalite (now grey gneiss) and 2865 ± 18 Ma Dachai granodiorite (Table 2.3). Igneous zircons from the Dachai granodiorite have $\varepsilon Hf_{(t)}$ values of +0.8 to +6.4 and $T_{DM}(Hf)$ model ages of 3.39−2.90 Ga (Table 2.3; Wu et al., 2013a), implying that the late Mesoarchean magmatism in the Qixia area was derived from partial melting of relatively young juvenile crustal rocks.

2.7.4 Neoarchean Geological Events

The Neoarchean events in the Eastern Block can be divided into the ~2.7 Ga event that formed the 2.75−2.65 Ga rock associations in the Luxi and Qixia areas and the ~2.5 Ga event that formed the 2.55−2.50 Ga rock associations over the whole Eastern Block. Although the 2.75−2.65 Ga rock associations are only exposed in the Luxi and Qixia areas (Table 2.1), the ~2.7 Ga event must have occurred across the whole Eastern Block and is a major crustal accretionary or mantle-extraction event that formed a thick mafic crust beneath the whole Eastern Block based on the following lines of evidence:

1. As given in Table 2.4, the 2.75−2.65 Ga rocks in the Luxi granite-greenstone terrane have positive $\varepsilon Hf_{(t)}$ values (+2.7 to +10.0), with most zircon Hf model ages close to the rock-forming ages (Table 2.4), which provides robust evidence that the ~2.7 Ga event that formed the 2.75−2.65 rock associations was a crustal accretion (mantle extraction) event (Jiang et al., 2010; Geng et al., 2012; Zhao and Zhai, 2013), not a crust-reworking event.
2. Also as given in Table 2.4, the 2.55−2.50 Ga rock associations in the Eastern Block possess mildly positive to slightly negative $\varepsilon Hf_{(t)}$ values, with most zircon Hf model ages pointing to 2.8−2.6 Ga, similar to rock-forming ages of the 2.75−2.65 Ga rock associations in the Luxi granite-greenstone terrane, suggesting that the 2.55−2.50 Ga rocks in the Eastern Block were mainly derived from the partial melting of an early Neoarchean (2.75−2.65 Ga) juvenile crust that formed at ~2.7 Ga. As the 2.55−2.50 Ga rocks are ubiquitous over the whole Eastern Block, the 2.7 Ga event must have occurred over the whole Eastern Block, forming an early Neoarchean juvenile crust that experienced partial melting or reworking to form the 2.55−2.50 Ga rock associations.

Sample No.	Rock Type	Location	Zircon Age (Ma)	εHf$_{(t)}$	Hf Model Age (Ma)	Sources
(1) 2.75–2.65 Ga Rock Associations						
Taishan Complex						
JN0742	Tonalite	Taishan	2691 ± 7	+4.13 to +6.41	2800–2733	Jiang et al. (2010)
M08	Garnet quartzite	Mengjiatun F.	2717 ± 33 2616 ± 19(m)	+3.75 to +7.26 +2.56 to +6.00	2848–2722 2830–2687	Du et al. (2010a)
D242	St–Grt mica schist	Mengjiatun F.	2742 ± 23 2642 ± 23(m)	+3.31 to +7.26 +4.03 to +7.07	2848–2731 2783–2674	Du et al. (2010a)
M03	Banded biotite gneiss	Mengjiatun F.	2695 ± 14	+3.21 to +6.27	2840–2738	Du et al. (2010a)
S0701	Hbl–Bi leptynite	Yanglingguan F.	2747 ± 7	+5.7 to +8.7	2821–2704	Wan et al. (2011a)
S0725	Biotite leptynite	Liuhang F.	2739 ± 16	+6.6 to +8.6	2780–2702	Wan et al. (2011a)
S0721	Biotite leptynite	Liuhang F.	2703 ± 6	+5.3 to +10.0	2797–2620	Wan et al. (2011a)
SY0311	Quartz diorite	E. Yanglingguan	2740 ± 6	+4.6 to +10.0	2854–2647	Wan et al. (2011a)
SY0336	Gneissic trondhjemite	Wangfushan	2711 ± 10	+2.7 to +5.1	2897–2797	Wan et al. (2011a)
S0503	Gneissic tonalite	Wangfushan	2714 ± 13 2640(m)	+4.8 to +8.3 +3.9 to +5.0	2823–2691 2975–2749	Wan et al. (2011a)
S0732	Gneissic trondhjemite	Duozhuang	2707 ± 9	+2.8 to +8.7	2893–2670	Wan et al. (2011a)
S0741	Gneissic granodiorite	Wangfushan	2712 ± 7	−13.6 to +2.1	3513–2924	Wan et al. (2011a)
Qixia Complex						
10SD18-1	Granodioritic gneiss	Nanfojia Village	2712 ± 12	−0.9 to +5.1	3490–2900	Wu et al. (2013b)
10SD21-1	Granodioritic gneiss	Huangyadi Village	2745 ± 12	−0.2 to +4.2	3390–3070	Wu et al. (2013b)
10SD20-1	Tonalitic gneiss	Hedongya Village	2710 ± 14	−2.0 to +1.1	3520–3260	Wu et al. (2013b)
QX19-1b	Monzogranitic gneiss	Yuke Town	2702 ± 51	+0.76 to +6.18	3025–2830	Liu et al. (2013b)

(*Continued*)

Table 2.4 (Continued)

Sample No.	Rock Type	Location	Zircon Age (Ma)	εHf$_{(t)}$	Hf Model Age (Ma)	Sources
QX44-1	Tonalitic gneiss	Xiyulin Village	2738 ± 23	+0.7 to +6.78	3.29−2.78	Liu et al. (2013b)
(2) 2.6−2.5 Ga Rock Associations						
Miyun						
FW04-120	TTG gneiss	SE Miyun	2521 ± 14	0 to +6.6	2824−2572	Yang et al. (2005)
Eastern Hebei						
FW04-28	Dioritic gneiss	Jiekouling	2528 ± 3	−0.3 to +4.4	2830−2670	Yang et al. (2008)
FW04-42	Granodioritic gneiss	Qinhuangdao	2522 ± 5 2497 ± 4(m)	−0.6 to +5.9 + 1.2 to +4.5	2870−2610 2770−2640	Yang et al. (2008)
FW04-54	Kf-granitic gneiss	Qinhuangdao	2523 ± 6 2451 ± 6(m)	+2.9 to +4.7 −1.1 to +2.3	2730−2660 2840−2690	Yang et al. (2008)
FW04-84	Granodioritic gneiss	Anziling	2524 ± 8 2490 ± 4(m)	+3.4 to +5.0 +1.1 to +3.8	2710−2640 2750−2660	Yang et al. (2008)
FW04-85	Tonalitic gneiss	Anziling	2522 ± 8 2501 ± 8(m)	−0.7 to +3.0 − 0.3 to +2.8	2860−2720 2710−2640	Yang et al. (2008)
04QA08	Olivine gabbro dyke	Qian'an	2516 ± 26	+1.9 to +4.2	2740−2668	Li et al. (2010a−c)
04QA08	Syenite dyke	Qian'an	2504 ± 11	+2.7 to +4.4	2705−2646	Li et al. (2010a−c)
XHG06	Amphibolite	Xihangou	2543 ± 9	−4.7 to +5.7	3030−2622	Lü et al. (2012)
WZZ06	Meta-felsic volcanic rock	Xihangou	2511 ± 12	−3.2 to +5	2928−2610	Lü et al. (2012)
11JD02-3	Granitic gneisses	Hanerzhuang	2614 ± 18	−1.4 to +9.0	2965−2635	Guo et al. (2013)
11JD12-1	Two-pyroxene granulite	Eastern Hebei	2523 ± 6	+2.5 to +6.5	2736−2622	Guo et al. (2013)
11JD17-2	Two-pyroxene granulite	Eastern Hebei	2609 ± 26	+0.9 to +5.8	2867−2684	Guo et al. (2013)
Jianping (Western Liaoning)						
YX05-1	Granodioritic gneiss	Jianping	2494 ± 18	+0.9 to +4.5	2770−2630	Wang et al. (2012a)

YX05-1	Dioritic gneiss	Jianping	2512 ± 185	+2.0 to +6.3	2740–2580	Wang et al. (2012a)
OCY31–2	Tonalitic gneiss	Jianping	2516 ± 6	+2.1 to +5.8	2747–2603	Wang et al. (2013b)
OCY33–1	Granodioritic gneiss	Jianping	2519 ± 11	+1.8 to +5.9	2757–2603	Wang et al. (2013b)
CY31–2	Tonalitic gneiss	Jianping	2513 ± 10	+4.0 to +10.1	2671–2437	Wang et al. (2013b)
CY46–1	Tonalitic gneiss	Jianping	2506 ± 12	+1.2 to +5.8	2767–2596	Wang et al. (2013b)
OCY37–1	Granodioritic gneiss	Jianping	2527 ± 17	+2.4 to +6.1	2746–2603	Wang et al. (2013b)
OCY46–1	Tonalitic gneiss	Jianping	2532 ± 7	+2.3 to +5.3	2749–2638	Wang et al. (2013b)
CY36–1	Monzogranitic gneiss	Jianping	2496 ± 12	+3.0 to +7.6	2691–2518	Wang et al. (2013b)
Fuxin-Suizhong (Western Liaoning)						
YX011-2	Andesitic (TTG?) gneiss	Fuxin	2567 ± 27	+2.7 to +6.2	2760–2630	Wang et al. (2011b)
FX009-1	Andesitic (TTG?) gneiss	Fuxin	2522 ± 8	+2.7 to +7.2	2730–2570	Wang et al. (2011b)
FX013-2	Andesitic (TTG?) gneiss	Fuxin	2589 ± 16	+3.3. to +8.3	2760–2570	Wang et al. (2011b)
Southern Liaoning (Within the Jiao–Liao–Ji Belt)						
DD09-2	Na-rich gneiss (TTG)	Dachangshan Island	2541 ± 9	−1.4 to +4.1	3030–2850	Meng et al. (2013a)
DD09-3	Na-rich gneiss (TTG)	Dachangshan Island	2537 ± 16	+2.9 to +7.2	2980–2630	Meng et al. (2013a)
DD09-6	Na-rich gneiss (TTG)	Dachangshan Island	2544 ± 14	+2.9 to +7.2	2930–2680	Meng et al. (2013a)
DD09-8	Na-rich gneiss (TTG)	Dachangshan Island	2541 ± 10	+0.3 to +5.1	3150–2730	Meng et al. (2013a)
DD29-1	Na-rich gneiss (TTG)	Guanglu Island	2544 ± 9	+0.6 to +4.2	3020–2790	Meng et al. (2013a)
DD13-1	Kf-granite	Changshan Island	2516 ± 11	+0.6 to +5.1	3060–2730	Meng et al. (2013a)
DD14-3	Kf-granite	Changshan Island	2514 ± 15	+0.9 to +8.6	2990–2510	Meng et al. (2013a)
DD14-4	Kf-granite	Changshan Island	2517 ± 17	+1.0 to +6.6	2950–2630	Meng et al. (2013a)

(Continued)

Table 2.4 (Continued)

Sample No.	Rock Type	Location	Zircon Age (Ma)	εHf(0)	Hf Model Age (Ma)	Sources
Yishui						
10SD07-2	Metapelitic	Beixiazhuang (Yishui)	2543 ± 6.3 2507 ± 9.4(m) +3.4	+2.2 to +4.1 2740	2820–2730	Wu et al. (2013a)
10SD07	Charnockite	Mashan pluton (Yishui)	2528 ± 8.3	+1.4 to +6.5	2860–2600	Wu et al. (2013a)
11SD01-2	Garnet charnockite	Caiyu pluton (Yishui)	2559 ± 16	+2.0 to +6.4	2920–2640	Wu et al. (2013a)
11SD02-16	Charnockite	Xueshan pluton (Yishui)	2535 ± 11	+2.1 to +7.0	2830–2660	Wu et al. (2013a)
10SD06-4	Granodioritic gneiss	Caiyu pluton (Yishui)	2570 ± 27 2499 ± 8(m)	+3.6 to +5.0 +1.5 to +1.2	2780–2710 2840–2830	Wu et al. (2013a)
YS06-19	Amphibolite	Yishui	2522 ± 5(m)	+3.2 to +7.6	2687–2522	Song et al. (2009)
YS06-41	Amphibolite	Yishui	2497 ± 4(m)	+2.9 to +5.7	2700–2592	Song et al. (2009)
YS06-40	Mafic granulite	Yishui	2514 ± 5(m)	+2.9 to +8.8	2703–2475	Song et al. (2009)
YS06-45	Mafic granulite	Yishui	2485 ± 10(m)	+1.2 to +7.5	2772–2525	Song et al. (2009)
YS06-49	Mafic granulite	Yishui	2493 ± 12(m)	+2.2 to +9.6	2725–2443	Song et al. (2009)
YS06-29	Amphibolite	Yishui	2531 ± 16	+1.5 to +5.9	2769–2591	Song et al. (2009)
YS06-30	Granitic gneiss	Lingshan pluton (Yishui)	2530 ± 7	−8.8 to +3.9	3140–2645	Song et al. (2009)
YS06-30	Granitic gneiss	Lingshan pluton (Yishui)	2530 ± 7	−8.8 to +3.9	3140–2645	Song et al. (2009)
YS06-31	Spinel-garnet granulite		2532–2537	+3.4 to +6.6	2709–2592	Song et al. (2009)
YS06-48	Granitic gneiss	Yishui	2531	+1.5 to +5.6	2760–2598	Song et al. (2009)
Luxi (Western Shandong)						
JN0740	Amphibolite	Taishan	2570 ± 18	+1.04 to +5.78	2768–2605	Jiang et al. (2010)
08YS-112	Monzogranite	Wanghailou (Feixian)	2543 ± 16	−1.06 to +5.67	3100–2680	Peng et al. (2013a)
08YS-105	Monzogranite	Hujiazhuang (Feixian)	2517 ± 21	+0.88 to +6.21	2960–2640	Peng et al. (2013a)
08YS-142	Monzogranite	Xiajiazhuan (Feixian)	2462 ± 18	−1.00 to +1.96	3040–2860	Peng et al. (2013a)

Qixia (Within the Paleoproterozoic Jiao–Liao–Ji Belt)

10SD10-2	Garnet amphibolite	3 km NW Qixia City	2546 ± 12 2458 ± 9(m)	+4.5 to +5.3 +2.5 to +5.5	2850–2700 2800–2620	Wu et al. (2013b)
10SD19-7	Garnet amphibolite	18 km SE Qixia City	2559 ± 10 2522 ± 11(m)	+2.1 to +7.7 +0.9 to +4.4	2880–2570 2930–2720	Wu et al. (2013b)
10SD27-9	Garnet amphibolite	5 km SE Qixia City	2473 ± 20(m)	−3.2 to +8.8	3130–2440	Wu et al. (2013b)
10SD16-1	Trondhjemitic gneiss	Daliujia Village	2526 ± 23 2501 ± 16	+4.1 to +7.0 +4.1 to +6.5	2870–2620 2860–2650	Wu et al. (2013b)
QX13-3	Tonalitic gneiss	Xiliu Town	2548 ± 12	+4.75 to 12.58	2800–2700	Liu et al. (2013b)
QX69-1	Granitoid gneiss	Yangjiaqiao Village	2522 ± 55	+3.43 to +8.02	2940–2520	Liu et al. (2013b)
QX27-2	Granitoid gneiss	Sunjiayanhou	2544 ± 15	+3.71 to +7.65	2900–2600	Liu et al. (2013b)
QX66-1	Tonalitic gneiss	Quanli Village	2564 ± 12	+3.00 to +10.71	3000–2500	Liu et al. (2013b)
Xenoliths in Phanerozoic Intrusions						
08JG18	Garnet-bearing gneiss	Jiagou intrusion	2552 ± 13	+2.3 to +3.9	2895–2798	Wang et al. (2012b)

Note: m = metamorphic age.

3. As most of the 2.55−2.50 Ga rock associations in the Eastern Block are TTG rocks, which are generally derived from the partial melting of a thickened mafic crust (eclogite or rutile-bearing amphibolite). This means that an early Neoarchean (2.75−2.65 Ga) juvenile crust formed by the ∼2.7 Ga event should be a mafic-dominant crust, which is either a lower continental crust or an oceanic crust. In any case, what can be inferred is that the ∼2.7 Ga event in the Eastern Block represented a Large Igneous Province (LIP) event that formed the main body of the Eastern Block, though most of this crust was reworked or destroyed by the late Neoarchean event at ∼2.5 Ga, except minor preserved in the Luxi and Qixia areas.

4. Nd isotopic data also support the conclusion that 2.8−2.6 Ga was the major crustal accretionary (or mantle extraction) period of the Eastern Block, whereas the ∼2.5 Ga geological event in the Eastern Block represents the reworking or recycling of the 2.8−2.6 Ga and even older crust (Wu et al., 2005b). As shown in Figure 2.13, the Nd isotopic characteristics of the 2.75−2.65 Ga TTG rocks and 2.8−2.7 Ga amphibolites are comparable, and they carry the mantle signature of positive $\varepsilon Nd(T)$ values, which suggests significant addition of juvenile crust in the period from 2.8 to 2.6 Ga (Jahn et al., 2008). The 2.55−2.50 Ga TTG rocks also have positive $\varepsilon Nd(T)$ values, similar to those of the 2.8−2.7 Ga amphibolites, suggesting that the former was most likely formed by partial melting of the latter. The ∼2.5 Ga granites possess negative $\varepsilon Nd(T)$ values, suggesting their derivation by remelting of older crustal material.

In summary, the ∼2.7 Ga magmatic event in the Eastern Block was a major crustal accretionary period, forming a thick mafic-dominant crust beneath the whole block, whereas the ∼2.50 Ga tecto-nomagmatic event was mainly involved in reworking or partial melting of the early Neoarchean mafic crust formed at ∼2.7 Ga, leading to the formation of numerous 2.55−2.50 Ga TTG rocks and minor 2.6−2.5 Ga volcanic rocks in the Eastern Block. This large-scale crustal reworking was terminated by a regional metamorphic event that happened over the whole Eastern Block at ∼2.5 Ga.

2.8 TECTONIC SETTINGS OF NEOARCHEAN GEOLOGICAL EVENTS

Controversy has long surrounded the tectonic settings of Archean crustal formation and evolution. At the center of the controversy is

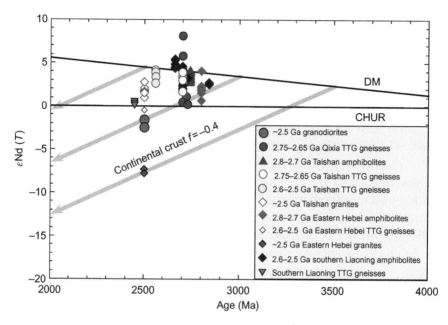

Figure 2.13 εNd(T) versus formation age (in Ga) plot for Neoarchean rocks from the Eastern Block of the NCC (Jahn et al., 2008). Data for three principal rock types are included: amphibolites, TTG gneisses, and granites. DM, depleted mantle evolution and CHUR, chondritic uniform reservoir evolution. The trend of isotope evolution for the continental crust is characterized by the average "f value" (=$^{147}Sm/^{144}Nd-1$) of -0.4 and is shown by the thick gray lines with arrow head. Data source: 2.75-2.65 Ga Qixia TTG gneisses (Jahn et al., 2008), 2.8-2.7 Ga amphibolites, 2.6-2.5 Ga TTG gneisses, and ~2.5 Ga granites in Taishan (Jahn et al., 1988); 2.8-2.7 Ga amphibolites and 2.6-2.5 Ga granitoids in Eastern Hebei (Jahn et al., 1987; Geng et al., 1999); 2.6-2.5 Ga TTG gneisses in Southern Liaoning (Wu et al., 1997).

whether or not Phanerozoic-style plate tectonics can be applied to the Archean. If not, what is the main tectonic mechanism that governed crustal formation and evolution during Archean time? To answer these questions, it is essential to carry out extensive field-based structural investigations followed by detailed metamorphic, geochemical, geochronological, and geophysical studies on various Archean terranes. As one of the oldest continental blocks in the world, the Eastern Block of the NCC is considered as a promising area for probing the tectonic settings and processes of Archean crustal formation and evolution because of excellent exposures of Eoarchean to Neoarchean basement rocks. However, as pre-Neoarchean basement rocks in the Eastern Block only make up ~5% of total exposure of the Archean basement, the available data for pre-Neoarchean rocks in the Eastern Block are insufficient to infer their tectonic environments and geodynamic processes. Therefore, this section only focuses on the tectonic settings of the Neoarchean crustal formation and evolution of the Eastern Block.

2.8.1 Tectonic Setting of the ~2.7 Ga LIP Event

As discussed above, the Neoarchean crustal formation and evolution of the Eastern Block were mainly involved in two major geological events: the ~2.7 Ga LIP event that formed a mafic crust in the Eastern Block and the ~2.5 Ga tectonothermal event that led to reworking or partial melting of the mafic crust or older crust to form the 2.55–2.50 Ga TTG rocks, granites, and volcanic rocks in the block. There is little argument on the tectonic setting of the ~2.7 Ga LIP event in the Eastern Block because it is generally considered that LIPs were formed by magmatism resulting from the decompressional melting of a mantle plume head, independently of plate tectonic setting (Campbell, 2005). This is also supported by geochemical data of ultra-mafic and mafic rocks from the Luxi greenstones that formed by the ~2.7 Ga LIP event. As mentioned earlier, the Luxi greenstones contain large volumes of komatiites and mafic rocks, of which the komatiites preserve typical spinifex textures (Figure 2.6), and are geochemically similar to other komatiites in the world. Polat et al. (2006) carried out detailed geochemical investigations on these komatiites, and their results indicate that they resulted from mantle plume activities. Later, Cheng and Kusky (2007) carried out further geochemical and isotopic work on the Luxi komatiites, and they also interpreted them to be the products of a ~2.7 Ga mantle plume. In fact, mantle plumes have been successfully applied to explain the formation of komatiites and associated mafic rocks from some other typical granite-greenstone belts in the Yilgarn Craton (Campbell et al., 1989; Hill et al., 1992) and the Superior Province of Canada (Tomlinson et al., 1998), because modern-style plate tectonics cannot well explain where and how MgO-rich komatiitic melts with eruption temperatures as high as 1650°C were generated (Hill et al., 1992). Considering geochemical characteristics of Archean komatiitic and basaltic rocks, and their close spatial and temporal association, Campbell et al. (1989) proposed that Archean komatiites were generated from hot, low-viscosity materials in the axial core of plumes, whereas the associated basalts were derived from the cooler head of the mantle plumes. Such a model has been well used to explain the origin of komatiites and mafic rocks from the Luxi greenstone terrane (Polat et al., 2006; Cheng and Kusky, 2007). Taken together, there is no doubt that the ~2.7 Ga LIP event in the Eastern Block resulted from an early Neoarchean mantle plume activity, though it is still unclear whether such a mantle plume event occurred in a continent setting or within an ocean environment. As pre-Neoarchean basement rocks in the Eastern Block only make up ~5% of total exposure of the Archean basement, it still remains controversial whether or not the whole Eastern Block had evolved into a matured continent by the early Neoarchean time when the ~2.7 Ga LIP event occurred. The absence of 2.75–2.65 Ga potassium-rich granites, which are generally considered to be generated from partial melting of a felsic crust, seems not to support the idea that the Eastern Block had been a matured continent by the time of the ~2.7 Ga LIP event. However, it is also possible that the continent of the Eastern Block was nearly all buried or assimilated by the continental flood basalts of the ~2.7 Ga LIP event. As estimated by Millard F. Coffin (http://www.ldeo.columbia.edu/~polsen/nbcp/lipmc. html), one million km^3 of basalt, the size of an average continental flood basalt province, would bury the area east of the Appalachians from Maine to Florida under more than a kilometer of basalt, which is larger than the area of the Eastern Block, nothing to say some super-LIPs like the Siberian trap. Alternatively, the ~2.7 Ga LIP event in the Eastern Block may have occurred in an oceanic setting in which the Anshan, Eastern Hebei, Luxi, and Qixia terranes existed as oceanic plateaus that were composed of continental crust (Puchtel et al., 1998). During Archean times, such oceanic plateaus may have existed forever in an oceanic basin if plate tectonics had not initialized by that time. No matter whether the ~2.7 Ga LIP event occurred in a continental setting or within an oceanic environment, it was a major crust-making event that formed a mafic-dominant crust beneath the whole Eastern Block, and then this mafic-dominant crust underwent extensive reworking (partial melting) during the ~2.5 Ga tectonomagmatic event, forming the 2.55–2.50 Ga TTG rocks, granites, and volcanic rocks in the Eastern Block.

2.8.2 Tectonic Setting of the ~2.5 Ga Tectonomagmatic Event

Controversy has long surrounded the tectonic setting of the ~2.5 Ga tectonomagmatic event in the Eastern Block, with one school of thought arguing for a magmatic arc environment under a plate tectonic regime (Zhang et al., 1980; Jahn and Zhang, 1984; Wang et al., 2012a; Liu et al., 2012d; Peng et al., 2012b), whereas another group of people favor a mantle plume model (Zhao et al., 1998, 1999b, 2001a; Geng et al., 2006, 2012; Yang et al., 2008; Wu et al., 2012). The magmatic arc model advocators argue that the Neoarchean TTG gneisses, occupying more than 85% of total exposure of Archean basement rocks in the Eastern Block, possess geochemical affinities to calc-alkaline rocks in modern magmatic arcs, which has led to a widely held hypothesis that the Neoarchean TTG gneisses in the Eastern Block represented the root zone of an Archean Andean-Type continental margin arc regime (Jahn and Zhang, 1984; Jahn et al., 2008; Wang et al., 2012a; Liu et al., 2012d; Peng et al., 2012b). Wang et al. (2012a) divided the 2.55–2.50 Ga TTG gneisses from the Jianping Complex into the high-Mg and low-Mg groups of which and the former was generated from the partial melting of a subducted oceanic slab, whereas the latter was derived from the partial melting of juvenile basaltic and pelitic rocks in the lower continental crust. Based on these data, Wang et al. (2012a) speculated that the Neoarchean TTG rocks from the Jianping Complex were formed in an Andean-type continental margin arc setting, which is also favored by Liu et al. (2012d) who argue that the 2.55–2.50 Ga mafic granulites, volcanic supracrustals, and TTG rocks in Western Liaoning record a transition from a mid-ocean ridge, through an intraoceanic island arc to an Andean-type active continental margin. Such a transitional process of plate tectonics regimes has also been applied to account for the tectonic settings of the 2.55–2.50 Ga TTG rocks in Eastern Hebei (Nutman et al., 2011), Northern Hebei (Wang et al., 2009a), and Western Shandong (Wan et al., 2010, 2011a; Peng et al., 2012b). Although such a magmatic arc model can well explain the arc-like geochemical features of the 2.55–2.50 Ga TTG rocks in the Eastern Block, it fails to accommodate the following geological characteristics:

1. The volume of 2.55–2.50 Ga granitoid (TTG) plutons is tremendous in the Eastern Block, covering 85% of the surface of the basement, but available zircon age data show no systematic age progression across the ~800 km wide (Kröner et al., 1998; Zhao et al., 1998; Geng et al., 2006; Grant et al., 2009; Wu et al., 2012), which is inconsistent with migrating or successively accreted magmatic arc models.

2. As mentioned earlier, the late Neoarchean greenstone terranes in the Eastern Block generally contain komatiites or komatiitic rocks, whereas the modern-style plate tectonics cannot well explain where and how MgO-rich komatiitic melts with eruption temperatures as high as 1650°C were generated in a continental margin arc system.

3. Most of the 2.6−2.5 Ga volcanic associations in the Eastern Block, like those in many other Neoarchean granite-greenstone terranes in the world, are bimodal, where basalt and ultramafic rocks are associated with dacite, rhyodacite, and rhyolite, without much andesite, which contrasts sharply with what we have observed in Phanerozoic continental margin arcs in which andesites are the dominant volcanic rock types (Hamilton, 1998).

4. An integration of rare-earth element (REE) features of TTG rocks and the results of modern petrological experiment precludes the possibility that cratonic-scale Archean TTG rocks were derived from partial melting of eclogites and its equivalents in subduction zones. Due to their high L/HREE ratios, high large-ion lithophile element contents, and negative Nb−Ta−Ti anomalies, Archean TTG rocks could be either derived from the partial melting of eclogites or rutile-bearing garnet amphibolites of subducted oceanic crust, or generated by melting the base of thick basaltic plateaus above mantle plumes, leaving behind restites containing pyroxene, garnet, and rutile (Bedard, 2006). However, modern petrological experiments have demonstrated that if TTG rocks were formed by partial melting of eclogites or garnet/rutile-bearing amphibolites, the partial melting degree should be lower than 30%; otherwise, the resultant granitoid rocks would not possess geochemical features of high L/HREE ratios, high large-ion lithophile element contents, and negative Nb−Ta−Ti anomalies that characterize the Archean TTG rocks. If the partial melting degree is no more higher than 30%, it requires that the actual volume of eclogites or garnet/rutile-bearing amphibolites that were partially melt to form the TTG rocks should be three times more than the volume of TTG rocks, which would create a mass-balance problem with the 2.55−2.50 Ga TTGs that make up 80% exposure of the Eastern Block, especially considering difficulty of extracting such a large volume of TTG melts from a subducting slab within such a short period (Zhao and Zhai, 2013).

5. The metamorphic $P−T$ paths of the late Neoarchean rocks in the Eastern Block are also inconsistent with those of subduction zones or magmatic arcs. As discussed earlier, the metamorphic evolution of the late Neoarchean rocks in the Eastern Block is characterized by anticlockwise $P−T$ paths involving IBC, which reflect the metamorphism related to the intrusion and underplating of large amounts of mantle-derived magmas. Although anticlockwise $P−T$

paths involving IBC may characterize the metamorphism occurring at the root of magmatic arcs or under back-arc basin setting (Wells, 1980; Bohlen, 1991), Brown (2006, 2007, 2008) argued that such metamorphism should be paired with the relatively high-pressure metamorphism of subducted zones that is characterized by clockwise $P-T$ paths involving isothermal decompression, and they together form the paired metamorphic belts like what we have observed in modern magmatic arcs. Brown (2008) argued that paired metamorphic belts could be regarded as a hallmark for tracing plate tectonics in the early history of the Earth. This is not the case in the Eastern Block where the ~ 2.5 Ga metamorphism is exclusively characterized by anticlockwise $P-T$ paths involving IBC (Figure 2.9), without any clockwise $P-T$ paths involving isothermal decompression to form paired metamorphic belts.

6. The dominant structural patterns of the Neoarchean lithotectonic elements in the Eastern Block are gneiss domes (dome-and-keel structures), which do not resemble linear orogenic belts that typify Proterozoic and Phanerozoic accretionary (subduction) orogens that resulted from metamorphosed and deformed continent margin arcs.

In summary, the late Neoarchean crust in the Eastern Block does not exhibit typical lithotectonic elements that are observed in Phanerozoic continental margin arcs. Considering the above dilemmas, Zhao et al. (1998) are the first to use a mantle plume model to explain the formation and evolution of the 2.55−2.50 Ga rocks in the Eastern Block. As summarized by Zhao et al. (1998, 1999b), a mantle plume model can well explain: (i) the exceptionally large exposure of TTG and granitoid intrusions that were emplaced over a short time period (2.55−2.50 Ga), without systematic age progression across a ~ 800 km wide block; (ii) the presence of MgO-rich komatiitic melts with eruption temperatures as high as 1650°C; (iii) bimodal volcanic assemblages; (iv) dominant diaprism-related domal structures; (v) metamorphism (with anticlockwise $P-T$ paths involving IBC) related to the intrusion and underplating of large amounts of mantle-derived magmas; and (vi) affinities of mafic rocks (metamorphosed to be amphibolites and mafic granulites) to continental tholeiitic basalts. Later, Zhao et al. (2001a) presented a detailed tectonic scenario discussing the upwelling of late Neoarchean mantle plumes beneath the Eastern Block and massive heat transfer resulting in widespread melting of the upper mantle and lower crust to form voluminous granitoid (TTG) magmas and the granulite-facies metamorphism with anti-clockwise $P-T$ paths involving IBC. Similar plume models have also been used to explain Neoarchean granite-greenstone terranes in Yilgarn Craton (Hill et al., 1989), Superior Craton (Bedard, 2006), and South India (Jayananda et al., 2000).

CHAPTER 3

Archean Geology of The Western Block

3.1 INTRODUCTION

Compared with the Eastern Block, the Western Block has not been intensely investigated until recently. In their initial threefold tectonic division of the North China Craton (NCC), Zhao et al. (1999a, 2001a) regarded the Western Block as a discrete continental block with uniform Neoarchean to Paleoproterozoic basement. Later, however, Zhao et al. (2002b, 2005) noticed that the Neoarchean and Paleoproterozoic basement rocks in the Western Block show contrasting structural and metamorphic features. For example, the Paleoproterozoic metamorphic rocks do not show a random distribution but are restricted to an east-west-trending linear tectonic belt, which is defined by strike-slip ductile shear zones, and the rocks in the belt are structurally characterized by large-scale thrusting and folding, transcurrent tectonics, sheath folds, and mineral lineations (Li et al., 1995a; Yang et al., 2000; Xu et al., 2001, 2002), most of which are classic indicators of collision tectonics. More importantly, the metamorphic evolution of the Paleoproterozoic rocks in the Western Block is characterized by clockwise $P-T$ paths involving nearly isothermal decompression (Jin, 1989; Jin et al., 1991, 1996; Lu, 1991; Lu et al., 1992, 1996; Liu et al., 1993; Lu and Jin, 1993; Zhao et al., 1999a), which is consistent with a continent—continent collisional environment, whereas the metamorphism of the Neoarchean basement rocks in the block is featured by anticlockwise $P-T$ paths involving isobaric cooling (Jin, 1989; Jin et al., 1991, 1996; Liu et al., 1993; Li et al., 1995a), which generally reflect an origin related to underplating of mantle-derived magmas (Zhao et al., 1999a and references wherein). Zhao et al. (1999a) previously considered the clockwise $P-T$ paths of Paleoproterozoic rocks from the Western Block as having resulted from the Paleoproterozoic collision between the Eastern and Western blocks. However, this model cannot reasonably explain the metamorphic evolution of the Paleoproterozoic rocks occurring far away from the Trans-North China Orogen, such as those in the Daqingshan, Wulashan, Qianlishan, and Helanshan complexes. For these reasons, Zhao et al. (2002a) proposed that the

Precambrian Evolution of the North China Craton. DOI: http://dx.doi.org/10.1016/B978-0-12-407227-5.00003-1

Paleoproterozoic rocks along an east-west-trending belt in the Western Block represent another Paleoproterozoic collision belt, along which the Yinshan Block represented by the late Archean basement in the north, and the Ordos Block covered by the Ordos Basin in the south, collided to form the Western Block in the Paleoproterozoic (Figure 3.1). Zhao et al. (2002a) used "Khondalite Belt" to name this belt because major lithologies in the belt are high-grade graphite-bearing pelitic granulites/gneisses, quartzites, calc-silicate rocks, and marbles, which are referred to as the "khondalite series" or the "Upper Wulashan Group" in the Chinese literature (Lu et al., 1992, 1996; Lu and Jin, 1993). In the Daqingshan and Wulashan areas, the khondalite series rocks were closely associated with late Neoarchean to Paleoproterozoic tonalite-trondhjemite-granodiorite (TTG) gneisses and mafic granulites, which were traditionally named the "Sangang Group" and the "Lower Wulashan Group" (Jin et al., 1991; Lu, 1991; Lu et al., 1992, 1996; Liu et al., 1993; Zhao et al., 2005). As most khondalite series rocks crop out marginal to the Ordos Basin, Zhao et al. (2002a) argued that they represent stable continental margin deposits in the continental margin basins surrounding the Ordos Block, whereas the TTG gneisses and mafic granulites that coexist with the khondalite series most likely represent a continental magmatic arc or island arc bordering the southern margin of the Yinshan Block

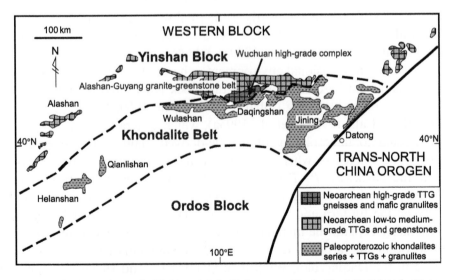

Figure 3.1 Sketch map showing the Yinshan and Ordos blocks separated by the Khondalite Belt in the Western Block (Zhao et al., 2002b, 2005).

(Liu et al., 1993). Later, Zhao et al. (2005) proposed that it was the Paleoproterozoic collision between the Yinshan and Ordos blocks that resulted in a spatial mixture of the khondalite series rocks, TTG gneisses and mafic granulites, constituting the Paleoproterozoic Khondalite Belt. The details of the Khondalite Belt will be discussed in Chapter 4. This chapter will focus on the Archean lithotectonic units in the Western Block.

3.2 ARCHEAN LITHOTECTONIC UNITS IN THE WESTERN BLOCK

Archean rocks in the Western Block are mainly distributed in the Yinshan Block, with minor amounts in the Daqingshan–Wulashan area of the Khondalite Belt, whereas the Ordos Block is entirely covered by Mesozoic and Cenozoic sedimentary rocks in the Ordos Basin (Figure 3.1), though available borehole and aeromagnetic data suggest the existence of granulite-facies basement rocks beneath the overlying Ordos Basin (Wu et al., 1986, 1998). The Yinshan Block is a typical Neoarchean block that can be further subdivided into the Alashan–Guyang granite-greenstone terrane in the north and the Wuchuan high-grade complex in the south (Figure 3.1).

3.2.1 Alashan–Guyang Granite-Greenstone Terrane

Some researchers have speculated that the Alashan (Alxa) terrane was not part of the Yinshan Block or event not part of the NCC, either because 900 − 700 Ma rocks, which are absent from the NCC but abundant in the Yangtze, Tarim, and Qaidam blocks, have been recognized at its southwestern margin in contact with the Northern Qilian Orogen (Geng et al., 2002; Zhang et al., 2012b), or because Neoarchean rocks like those in the Yinshan Block were not found in the Alashan terrane (Dan et al., 2012). For example, Li (2004) regarded the Alashan (Alax) and Dunhuang terranes as an independent block, not part of the NCC. Ge and Liu (2000) argued that the Alashan terrane was part of the so-called "Western China Craton" including the Tarim, Qaidam, and Alashan blocks that contain abundant early Neoproterozoic (900 − 750 Ma) rocks. Based on the absence of Archean rocks and zircons in the terrane, Dan et al. (2012) and Zhang et al. (2012b) also proposed that the Alashan terrane was not the western extension of the Western Block. However, all of these models were established on lithological differences between the

Alashan terrane and other parts of the NCC, but none has provided convincing evidence for the existence of a continent–continent collisional belt between the Alashan terrane and the NCC. As discussed in Chapter 1, recognition of a continent–continent collisional belt is the key to distinguishing continental blocks, and as long as two terranes are separated by a continent–continent collisional belt, they should be regarded as two discrete continental blocks no matter whether their lithotectonic features are similar or different. For this reason, in this book the Alashan terrane is considered as part of the NCC, representing the western extension of the Yinshan Block. This is supported by recent discovery of ~2.52 Ga banded TTG gneisses in the Alashan terrane (Gong et al., 2012), whose rock-forming and metamorphic ages and zircon Hf data are remarkably similar to those of TTG rocks in the Yinshan Block (Table 3.1). As data for Neoarchean rocks in the Alashan area are extremely limited, the following discussion will be focused on the Guyang granite-greenstone belt.

The Guyang granite-greenstone belt extends from Mount Serteng in the west to Donghongsheng in the east (Figure 3.2), of which the former exposes a complete sequence of greenstones. The greenstones can be further subdivided into three units, of which the lower unit is dominated by ultramafic (komatiitic) and mafic volcanic rocks metamorphosed from greenschist- to amphibolite-facies, including serpentinite, serpentinized actinolitite, tremolitite, greenschist, amphibolites, and hornblende-plagioclase gneiss, with minor calc-alkaline volcanic rocks and banded iron formations (Li et al., 1987, 1995a; Ma et al., 2010, 2013a,b,c; Jian et al., 2012; Liu et al., 2012f). The middle unit consists predominantly of basaltic and felsic volcanics and volcanoclastic rocks, represented by greenschist, amphibolites, and hornblende-plagioclase gneiss (Li et al., 1995a), and the upper unit is dominated by sedimentary rocks metamorphosed from greenschist- to amphibolite-facies, forming paragneiss, mica schist, quartzite, and marble (Li et al., 1995a). In the lower unit, serpentinite, serpentinized actinolitite and tremolite are komatiitic in composition, geochemically similar to those komatiites in other greenstone terranes in the world, though they do not preserve typical spinifex textures (Chen, 2007; Ma et al., 2010). Some of these komatiitic rocks contain orthopyroxene and anthophyllite, which show a preferred orientation parallel to the regional foliation (schistosity). Based on geochemical features, Ma et al. (2010) divided the komatiitic rocks into two types (the Munro and Barberton

Table 3.1 Summary of Zircon Ages of Archean Rocks in the Yinshan Block

Sample	Rock Name	Location	Rock-Forming Age (Ma)	Metamorphic Age (Ma)	εHf(0)/TDM(Hf) Age	Method	Sources
Alashan—Guyang Granite-Greenstone Terrane							
LS10-8	Banded gneiss	Alashanyouqi	2522 ± 30	2496 ± 11	0.4 to 4.9/2.7–3.0 Ga	SHRIMP	Gong et al. (2012)
GY50	Andesite	Guyang	2510 ± 7			SHRIMP	Jian et al. (2012)
GY50-1	Andesitic dacite	Guyang	2515 ± 10			LA-ICP-MS	Chen (2007)
GY50-2	Basalt	Guyang	2516 ± 10			LA-ICP-MS	Chen (2007)
SHM-2	Amphibolite (basalt)	Sanheming	2562 ± 14			Cameca	Liu et al. (2012f)
08XM04	Quartz diorite	Banshentu pluton	2480 ± 20		−2.1 to −2.1/2.9–2.7 Ga	LA-ICP-MS	Ma et al. (2013a)
GY32	Trondhjemite	Xiwulanbulang	2520 ± 9			SHRIMP	Jian et al. (2005)
GY38-1	Tonalite	Hejiao	2516 ± 7			SHRIMP	Jian et al. (2012)
GY37-1	Potassium granitic layer	Hejiao	2515 ± 9			SHRIMP	Jian et al. (2012)
GY51-2	Pegmatite	Guyang	2508 ± 17			SHRIMP	Jian et al. (2012)
GY51-1	High-Mg andesite	Guyang	2533 ± 5			SHRIMP	Jian et al. (2012)
Unknown	TTG	Guyang	2534 ± 7			SHRIMP	Ren (2010)
B4206-1	Gneissic granite	Hongshanzi	2509 ± 7		+0.9 to +6.1/2.8 − 2.6 Ga	SHRIMP	Dong (2012)
2P3b14-2	Quartz diorite	Hongshanzi	2534 ± 7	2483 ± 20	−6.6 to +4.4/2.8 − 2.6 Ga	SHRIMP	Dong (2012)
4P9Gs37	Quartz diorite	Qianhao	2535 ± 16			SGD	Zhang (2004)

(Continued)

Table 3.1 (Continued)

Sample	Rock Name	Location	Rock-Forming Age (Ma)	Metamorphic Age (Ma)	$\varepsilon Hf_{(t)}/T_{DM}(Hf)$ Age	Method	Sources
Wuchuan High-Grade Complex							
B1001	Gneissic trondhjemite	Xiwulanbulang	2697 ± 11	2539 ± 9	+1.8 to +8.3/2.8 – 2.6 Ga	SHRIMP	Dong et al. (2012b)
NM1132	Gneissic trondhjemite	Xiwulanbulang	2692 ± 17	2528 ± 16	−0.12 to +8.2/ 2.8 – 2.6 Ga	SHRIMP	Ma et al. (2013d)
B9007	Felsic granulite	Housaierdong	2545 ± 10	2503 ± 10	+1.0 to +6.5/2.8 – 2.6 Ga	SHRIMP	Dong et al. (2012a)
B9009	Felsic granulite	Xiwulanbulang	2516 ± 15	2472 ± 14	+1.0 to +8. 6/2.8 – 2.6 Ga	SHRIMP	Dong et al. (2012a)
B9008	Doritic granulite	Housaierdong	2506 ± 9	2499 ± 22	+1.6 to +8. 1/2.9 – 2.5 Ga	SHRIMP	Dong et al. (2012a)
2P7B40-1	Garnet charnockite	Xiwulanbulang	2547 ± 12	2497 ± 19	−4.2 to +7.4/2.8 – 2.7 Ga	SHRIMP	Dong (2012)
XWLBL01	Garnet gneiss	Xiwulanbulang	2521 ± 6	2515 ± 18		SHRIMP	Jian et al. (2012)
XWLBL02	Mafic granulite	Xiwulanbulang	2544 ± 5	2503 ± 12		SHRIMP	Jian et al. (2012)
XWLBL03	Enderbite	Xiwulanbulang	2525 ± 3			SHRIMP	Jian et al. (2012)
GY49	diorite (sanukitoid)	Xiwulanbulang	2556 ± 14			SHRIMP	Jian et al. (2005)
ZLG04	Amphibole gneiss	Zhulagou	2526 ± 5	2536 – 2321		SHRIMP	Jian et al. (2012)
ZLG05	Charnockite	Zhulagou	2535 ± 5			SHRIMP	Jian et al. (2012)
ZLG02	diorite	Zhulagou	2520 ± 6			SHRIMP	Jian et al. (2012)
ZLG03	Granitic xenolith	Zhulagou	2594 ± 5			SHRIMP	Jian et al. (2012)
ZH01	Trondhjemite	Zhaohe	2502 ± 6			SHRIMP	Jian et al. (2012)

08XM20	Diorite (sanukitoid)	Dajitu pluton	2523 ± 13		+1.2 to +2.1/2.8–2.6 Ga	LA-ICP-MS	Ma et al. (2013a)
08XM15	Tonalitic gneiss	Haolagou pluton	2465 ± 18		+0.4 to +5.4/2.8–2.6 Ga	LA-ICP-MS	Ma et al. (2013a)
08XM36	Dioritic charnockite	Wuchuan	2524 ± 4	2498 ± 3	+1.4 to +4.9/2.8–2.6 Ga	Cameca	Ma et al. (2013b)
08XM40	Tonalitic charnockite	Wuchuan	2533 ± 15	2490 ± 11	+2.3 to +5.5/2.7–2.6 Ga	LA-ICP-MS	Ma et al. (2013b)
1P9TW22	Charnockite	Xiwulanbulang	2511 ± 5			SGD	Zhang et al. (2003)
Unknown	felsic granulite	Xiwulanbulang	2509 ± 7			SHRIMP	Ren (2010)
Unknown	Felsic granulite	Xiwulanbulang	2483 ± 20			SHRIMP	Ren (2010)

SGD, single grain dissolution; SHRIMP, sensitive-high resolution ion microprobe.

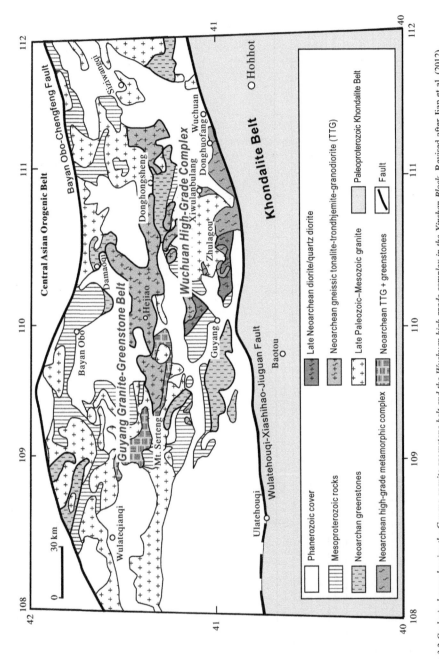

Figure 3.2 *Geological map showing the Guyang granite-greenstone belt and the Wuchuan high-grade complex in the Yinshan Block. Revised after Jian et al. (2012).*

types), of which the first type is characterized by relatively high MgO (29 − 26%) and TiO_2 (0.31 − 0.24%) but low Al_2O_3 (4.8 − 4.2%), with Al_2O_3/TiO_2 ratios ranging between 20 and 10, $(La/Yb)_N$ of 1.6−1.8, and $(Gd/Yb)_N$ of 0.7−0.8, whereas the second type is characterized by low MgO ($\sim$25.5%) but high Al_2O_3 (8.4 − 6.0%) and TiO_2 (0.23 − 0.19%), with Al_2O_3/TiO_2 ratios ranging between 44 and 30, $(La/Yb)_N$ of 1.4−2.2, and $(Gd/Yb)_N$ of 1.6 − 1.1. In the Al_2O_3−MgO−$(Fe_2O_3^T + TiO^2)$ diagram, both types of the komatiitic rocks plot in the komatiite field (Ma et al., 2010). Using the laser ablation-induced coupled plasma-mass spectrometry (LA-ICP-MS) dating technique, Chen (2007) obtained $^{207}Pb/^{206}Pb$ ages of 2516 ± 10 Ma and 2515 ± 10 Ma for meta-basalt (amphibolite) and meta-dacite, respectively, from the Guyang greenstone sequences, which are interpreted as the timing of volcanism to have formed these rocks. Using the secondary ion mass spectrometry (SIMS) dating techniques, Jian et al. (2012) obtained a $^{207}Pb/^{206}Pb$ age of 2510 ± 7 Ma for andesite, and Liu et al. (2012f) obtained a $^{207}Pb/^{206}Pb$ age of 2562 ± 14 for meta-basalt (amphibolite) from the lower sequence of the Guyang greenstones. Taken together, these data demonstrate that the Guyang greenstone volcanism occurred in the period 2560 − 2510 Ma.

Granitoid rocks from the Guyang granite-greenstone terrane are typical TTG suites that are dominated by tonalite and granodiorite with minor trondhjemite, quartz diorite, and diorite metamorphosed in greenschist- and amphibolite-facies, of which some quartz diorite and diorite are considered to be adakite (Zhang et al., 2006b) and sanukitoids (Jian et al., 2005; Ma et al., 2013a), mainly occurring along the boundary between the Yinshan Block and the Khondalite Belt (Figure 3.2). According to Li et al. (1995a), tonalite, trondhjemite, and granodiorite do not show intrusive relationships but are gradational, and locally they intrude the greenstone sequences, although in most cases they are in tectonic contact with the latter. These rocks, together with associated greenstones, were previously considered to have formed during Mesoarchean time (Qian et al., 1985) or even in the Paleoarchean (Yang et al., 2000), but recent SIMS and LA-ICP-MS zircon dating results show that they were emplaced in the late Neoarchean (2.6 − 2.5 Ga), mostly at 2.55 − 2.50 Ga (Table 3.1). Like those TTGs in other granite-greenstone terranes in the world, the TTG rocks from the Guyang granite-greenstone terrane are geochemically characterized by high Sr and Ba contents, low Y and Yb contents, and high Sr/Y ratios (Dong, 2012; Ma et al., 2013a). They are significantly enriched in LREE and large-ion lithophile elements (LILEs) such as Th, Ba, and K, strongly depleted in HREE, with $(La/Yb)_N$ ratios of 27–76, strong positive Eu anomalies, and weakly positive Sr and Ba anomalies (Ma et al., 2013a). These geochemical features indicate that their parental melts originated from partial melting of hydrated basaltic rocks at high pressures, with garnet, but no plagioclase, in the residue. Generally, these geochemical features can be interpreted as a result of partial melting of hydrated basaltic rocks under high-pressure conditions with garnet, rutile, and amphibole but no plagioclase in the residual phases (Rapp et al., 1991; Springer and Seck, 1997); although the tectonic settings in which the basaltic rocks are formed and subsequently partially melted still remain the subject of debate, with one school of thought favoring basaltic material that was previously underplated beneath thickened crust (Smithies 2000; Smithies et al. 2003), most probably through mantle plumes (Zhao et al., 1998, 1999b, 2001a), whereas others argue for basaltic rocks in subducting oceanic slab that melted in a hotter Archean mantle (Martin et al. 2005). In addition, Bedard (2006) suggested that when a hydrous basaltic lower crust was delaminated into the mantle, it could be partially melted to produce TTGs (see discussion later).

3.2.2 Wuchuan High-Grade Complex

The Wuchuan high-grade complex occupies a small area in the southern part of the Yinshan Block, extending from Zhulagou, through Donghuofang, to Xiwulanbulang, and comprises Neoarchean granitoid plutons, granulites, and charnockitic rocks (Figure 3.3). The Neoarchean granitoid plutons are dominated by high- to medium-grade metamorphosed diorite or quartz diorite, which are interpreted as the same sanukitoid suite as those in the Guyang greenstone belt (Jian et al., 2005; Ma et al., 2013a). The granulites from the Wuchuan high-grade complex represent the mixtures of felsic granulites dominated by hypersthene-bearing dioritic and tonalitic gneisses (high-grade metamorphosed TTG rocks) and granulite-facies supracrustals that include mafic granulites (garnet two-pyroxene granulite), garnet-pyroxene-bearing banded iron formations, and minor retrograded amphibolites, which occur as sheets, boudins, and elongate lenses within the felsic granulites. The charnockitic rocks are weakly gneissic to massive granitoid rocks, commonly containing different contents of K-feldspar, and are considered as the anatectic products of the felsic

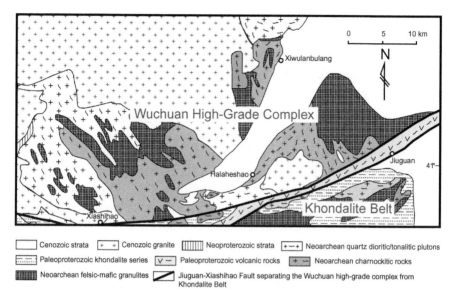

Cenozoic strata Cenozoic granite Neoproterozoic strata Neoarchean quartz dioritic/tonalitic plutons
Paleoproterozoic khondalite series Paleoproterozoic volcanic rocks Neoarchean charnockitic rocks
Neoarchean felsic-mafic granulites Jiuguan-Xiashihao Fault separating the Wuchuan high-grade complex from Khondalite Belt

Figure 3.3 Major lithotectonic units in the Wuchuan high-grade complex. After Dong (2012).

granulites (hypersthene-bearing dioritic and tonalitic gneisses) under granulite-facies metamorphism (Li et al., 1987, 1995a; Jin, 1989; Jin et al., 1991). This is supported by their relationships in the field where the charnockitic rocks are transitional from the charnockitic gneisses with minor amounts of K-feldspar, through gneissic charnockites with moderate amounts of K-feldspar, to massive K-charnockites with large amounts of K-feldspar, of which the charnockitic gneisses always show gradational relationships with the felsic granulites, interpreted as the products of *in situ* anatexism of the latter, whereas in most cases the gneissic charnockites intrude the felsic granulites, but in some places they are in gradational contacts. In contrast, the massive charnockitic bodies are always intrusive into the felsic granulites on their boundaries and contain a large amount of felsic granulite xenoliths within the bodies. The foliations of felsic granulite xenoliths within the massive charnockites are not consistent but random, suggesting that they were rotated during the ascent of the charnockitic magmas.

Microscopic observations show that the charnockitic rocks contain both the residual and crystallization phases of minerals, of which the residual phases are anhedral hypersthene, clinopyroxene, garnet, plagioclase, and quartz, whereas the crystallization phases are subhedral K-feldspar, plagioclase biotite, and quartz (Dong, 2012; Dong et al.,

2012a). The proportions of the crystallization phases to residual phases are different among different types of the charnockites, of which the massive charnockites contain more crystallization phases (subhedral K-feldspar, plagioclase biotite, and quartz) than the charnockitic gneisses and gneissic charnockites.

Geochemically, the felsic granulites, that is, hypersthene-bearing dioritic and tonalitic gneisses, are characterized by high Na_2O (3.21−5.08 wt%) and Al_2O_3 (12.25−16.68 wt%), negative Nb and Ti anomalies, and REE patterns with high LREE enrichment and no negative europium anomaly (Li et al., 1995a; Dong, 2012; Ma et al., 2013b), similar to those of TTG rocks in the Guyang granite-greenstone belt. The charnockitic gneisses and gneissic charnockites are geochemically similar to the felsic granulites, though the former are high in K_2O, whereas the massive charnockites are geochemically different from the felsic granulites in that the former contain more SiO_2, K_2O, and Na_2O but less FeO, MgO, and CaO (Dong, 2012; Ma et al., 2013b). In addition, the massive charnockites are rich in large-ion incompatible elements of K, Rb, Ta, and Ba, with a high Rb/Sr ratio (Dong, 2012), which support their anatectic origin. In terms of rare earth elements (REE), the massive charnockites are higher in LREE/HREE (($La/Yb)_N > 100$) than charnockitic gneisses and gneissic charnockites, which have REE features similar to those of the felsic granulites.

As given in Table 3.1, except minor trondhjemitic gneisses that formed at ∼2.7 Ga, most rocks from the Wuchuan high-grade complex formed at 2.55 − 2.50 Ga, coevally with those rocks in the Guyang granite-greenstone belt. This led some researchers to have proposed that the former were the lower crustal equivalents of the latter (Li et al., 1995a), that is, the Guyang granite-greenstone belt represented the upper crust that experienced greenschist- to lower amphibolite-facies metamorphism, whereas the Wuchuan high-grade complex represented the lower crust that underwent upper amphibolite- to granulite-facies metamorphism in association with extensive anatexis to have formed voluminous *in situ* or subautochthonous charnockitic migmatites and massive charnockites. In addition, both the charnockitic rocks and mafic−felsic granulites from the Wuchuan high-grade complex have metamorphic zircons that occur either as overgrowth rims surrounding older magmatic zircon cores or as single grains, which are structureless, highly luminescent and with very low Th/U ratios, making them distinctly different from the magmatic zircons that are generally characterized by oscillatory zoning, low luminescence, and comparatively high Th/U ratios. As given in Table 3.1, metamorphic zircons from both the charnockitic rocks and mafic–felsic granulites yield similar concordant $^{207}Pb/^{206}Pb$ ages around 2.5 Ga, indicating that the Wuchuan high-grade complex underwent a regional granulite-facies metamorphic event at the end of Archean. Moreover, like those Neoarchean basement rocks in the Eastern Block, the $2.55 - 2.50$ Ga granitoid (TTG) gneisses in the Wuchuan high-grade complex are also characterized by near-contemporaneity of granitoid intrusive and metamorphic events, with the peak of metamorphism occurring shortly after the widespread intrusion of granitoid suites (Table 3.1).

3.3 METAMORPHIC EVOLUTION

Available data show that the metamorphic evolution of the Neoarchean basement rocks from both the Wuchuan high-grade complex and the Guyang granite-greenstone belt is characterized by an anticlockwise $P-T$ path involving near-isobaric cooling (Figure 3.4), which also characterizes the metamorphic evolution of those Neoarchean basement rocks in the Daqingshan–Wulashan area of the Paleoproterozoic Khondalite Belt.

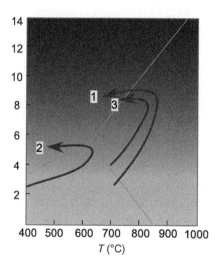

Figure 3.4 P–T *paths of Neoarchean basement rocks in the Yinshan Block: 1, Wuchuan high-grade complex (Jin et al., 1991; Lu and Xu, 1995); 2, Guyang granite-greenstone belt (Jin, 1989); 3, Neoarchean rocks in the Daqingshan–Wulashan area of the Khondalite Belt (Jin et al., 1991; Liu et al., 1993).*

In the Wuchuan high-grade complex, Jin et al. (1991) and Lu and Xu (1995) recognized four distinct metamorphic mineral assemblages (M1 to M4) from the mafic granulites, of which M1 is an amphibolite-facies assemblage of plagioclase + hornblende ± quartz occurring as mineral inclusions within garnet, clinopyroxene, orthopyroxene grains. The temperatures of M1 were estimated at 680–750°C by using the plagioclase–hornblende geothermometry (Jin, 1989). M2 is a medium-pressure granulite-facies assemblage of garnet + orthopyroxene + clinopyroxene + plagioclase ± quartz, and its metamorphic $P-T$ conditions were estimated by the garnet–orthopyroxene–clinopyroxene–plagioclase–quartz geobarometers (Newton and Perkins, 1982; Perkins and Chippera, 1985) and the two-pyroxene (Bohlen and Essene, 1979), garnet–clinopyroxene/orthopyroxene geothermometers (Ellis and Green, 1979; Sen and Bhattacharya, 1984; Ganguly et al., 1988) at 8.0–10.0 kbar and 800–900°C (Jin, 1989; Lu and Xu, 1995). The M3 assemblage is characterized by garnet + quartz or garnet + clinopyroxene symplectic coronas surrounding orthopyroxene, clinopyroxene, and plagioclase grains (Figure 3.5), which is called the "red-eye-socket" texture in the Chinese literature and contrasts with the "white-eye-socket" texture assigned to the plagioclase + orthopyroxene symplectic coronas surrounding garnet grains. The $P-T$ conditions of this assemblage are 9.0–10.0 kbar and 700–750°C, estimated from garnet–orthopyroxene–clinopyroxene–plagioclase–quartz geobarometers (Newton and Perkins, 1982; Perkins and Chippera, 1985) and garnet–clinopyroxene/orthopyroxene thermometers (Ellis and Green, 1979; Sen and Bhattacharya, 1984; Ganguly et al., 1988). The M4 assemblage is represented by cummingtonite + plagioclase replacing clinopyroxene and orthopyroxene, forming in amphibolite-facies, but its $P-T$ cannot be estimated because of the lack of suitable thermobarometers. These mineral assemblages and their $P-T$ conditions define an anticlockwise $P-T$ path involving near-isobaric cooling for the Wuchuan high-grade complex (Figure 3.4).

For the Guyang granite-greenstone belt, the estimations of pressure and temperature of metamorphism from mineral equilibria are difficult due to the lack of suitable calibrated geobarometers and geothermometers for low-grade greenstone rocks. The best data for its metamorphic evolution were obtained from minor lower amphibolite-facies rocks from the Guyang granite-greenstone belt. Jin (1989) recognized three metamorphic mineral assemblages (M1, M2, and M3) from

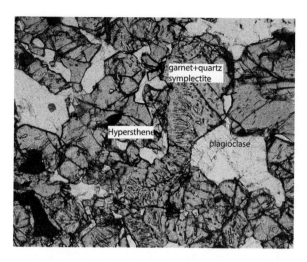

Figure 3.5 Garnet + quartz symplectic coronas surrounding orthopyroxene, clinopyroxene, and plagioclase grains.

garnetiferous amphibolites, of which the M1 assemblage is a typical greenschist-facies assemblage of actinolite + chlorite + epidote + Na-rich plagioclase, preserved as inclusions within garnet grains. Generally, the temperature of such a greenschist-facies assemblage ranges from 350°C to 450°C, but its pressure has a wide range, from 2.0 to 6.0 kbar (Mason, 1990). The M2 assemblage is hornblende (core) + plagioclase (core, An30) + quartz ± garnet, forming at the peak metamorphic stage. The $P-T$ conditions of the M2 assemblage were estimated by the plagioclase–hornblende thermobarometer and the garnet–hornblende thermometer at ca. 600–650°C and 5.0–6.0 kbar (Jin, 1989). The M3 assemblage is represented by retrogressive hornblende (rim) + epidote + plagioclase (rim, An20) + quartz, and its $P-T$ conditions were estimated by the hornblende–plagioclase thermobarometer at 5.0 kbar and 450°C (Jin, 1989). The transition from the M2 to M3 stage reflects a near-isobaric cooling process. These mineral assemblages and their $P-T$ conditions define an anti-clockwise $P-T$ path near-isobaric cooling for the Guyang granite-greenstone belt (Figure 3.4; Jin, 1989), though it remains questionable as the pressure of M1 has not been determined.

As mentioned above, minor Neoarchean rocks are exposed in the Daqingshan–Wulashan area of the Khondalite Belt (Figure 3.5). They are composed primarily of Neoarchean high-grade TTG gneisses and mafic granulites that are associated with Paleoproterozoic khondalite series rocks (Jin et al., 1991; Liu et al., 1993). The metamorphic evolution of the Neoarchean mafic granulites and high-grade TTG gneisses in the Daqingshan—Wulashan area has been described by a number of authors (Jin, 1989; Jin et al., 1991; Liu et al., 1993; Lu, 1991; Lu and Xu, 1995; Lu and Jin, 1993; Jin and Li, 1994, 1996). According to Jin et al. (1991) and Liu et al. (1993), the mafic granulites and high-grade TTG gneisses from the Daqingshan—Wulashan display three metamorphic assemblages, of which the M1 assemblage includes hornblende, plagioclase, and quartz, which occur as inclusions within coarse-grained garnet, clinopyroxene, and orthopyroxene grains. In most cases, these inclusion-type minerals occur as single grains enclosed in the host minerals, but in a few cases, they appear together within orthopyroxene, clinopyroxene, and garnet grains. The temperature of the M1 assemblage was estimated at 700—750°C, based on compositions of the inclusion-type hornblende and plagioclase analyzed by Liu et al. (1993), using the hornblende—plagioclase thermometer of Holland and Blundy (1994). The pressure of M1 was estimated by Liu et al. (1993) at 3.8—4.2 kbar by using the plagioclase—hornblende thermobarometer of Plyusnina (1982), but this result is doubtful as the thermobarometer they used can be applied only for the epidote amphibolite-facies assemblages (Plyusnina, 1982). The peak (M2) assemblage is coarse-garnet + clinopyroxene + hypersthene + plagioclase ± quartz ± hornblende. In many cases, these minerals are in contact with straight grain boundaries and ~120° triple junctions, indicating that they have reached equilibrium during the peak metamorphism. The $P-T$ conditions of the M2 assemblage were estimated at 8.7—10.5 kbar and 800—850°C using garnet—orthopyroxene—clinopyroxene—plagioclase—quartz geobarometers (Newton and Perkins, 1982; Perkins and Chippera, 1985) and two-pyroxene (Bohlen and Essene, 1979) and garnet—clinopyroxene (Ellis and Green, 1979; Ganguly et al., 1988) geothermometers. The post-peak (M3) assemblage is represented by the garnet + quartz ± clinopyroxene symplectic coronas around the peak minerals of plagioclase, clinopyroxene, and orthopyroxene, that is, the "red-eye-socket" texture, which has been described by Jin (1989), Jin et al. (1991), Liu et al. (1993), and Jin and Li (1994, 1996). The $P-T$ conditions of M3 were estimated at 8.7—10.5 kbar and 770—810°C using the garnet—pyroxene—plagioclase—quartz geobarometers of Newton and Perkins (1982) and Perkins and Chippera (1985) and garnet—clinopyroxene geothermometers of Ellis and Green (1979) and Ganguly et al.

(1988). These $P-T$ estimates indicate an isobaric cooling process from M2 to M3. Although there are some uncertainties with the pressure estimation for M1, the $P-T$ estimates for the M1, M2, and M3 assemblages define anticlockwise $P-T$ path involving isobaric cooling (IBC) for the Neoarchean mafic granulites and high-grade TTG gneisses in the Daqingshan–Wulashan area of the Khondalite Belt (Figure 3.4, Liu et al., 1993), which contrasts with the clockwise $P-T$ paths involving isothermal decompression reconstructed for the khondalite series rocks that are tectonically associated with the mafic granulites and high-grade TTG gneisses in the belt (see discussion in Chapter 4).

Taken together, the Neoarchean basement rocks in the Western Block exhibit IBC-type anticlockwise $P-T$ paths, which reflect an origin related to the intrusion and underplating of large amounts of mantle-derived magmas that not only provide heat for the metamorphism but also add a large volume of mostly mafic material to the base of the crust. Crustal growth is therefore dominated by vertical addition of mantle-derived magmas to crust.

3.4 MAJOR GEOLOGICAL EVENTS AND THEIR TECTONIC SETTINGS

3.4.1 Major Geological Events

Unlike the Eastern Block that underwent multiple and complicated geologic events from the Eoarchean, through Paleoarchean and Mesoarchean, to Neoarchean times, the Western Block (Yinshan Block) only experienced Neoarchean crustal accretion and reworking. As given in Table 3.1, except minor ~ 2.7 Ga trondhjemite exposed in the Xiwulanbulang area (Dong et al., 2012b; Ma et al., 2013d), all other rocks, including TTG suites and ultramafic–mafic volcanic rocks in the Guyang granite-greenstone belt and the Wuchuan high-grade complex, formed within a short period between 2.55 and 2.50 Ga. However, zircon Hf model ages of the $2.55 - 2.50$ Ga rocks have revealed that the ~ 2.7 Ga event must have been a major crustal accretionary or mantle-extraction event that formed a thick mafic crust beneath the Yinshan Block. As given in Table 3.1, most of the $2.55 - 2.50$ Ga rocks from the Yinshan Block possess positive $\varepsilon Hf_{(t)}$ values, with zircon Hf model ages pointing to $2.8 - 2.6$ Ga, similar to rock-forming age of the ~ 2.7 Ga trondhjemite exposed in the Xiwulanbulang area, suggesting that the $2.55 - 2.50$ Ga rocks in the

Yinshan Block were mainly derived from the partial melting of a juvenile crust that formed at ~ 2.7 Ga. This is also supported by positive $\varepsilon Nd_{(t)}$ values obtained for the charnockitic rocks and amphibolites from the Wuchuan high-grade complex and the Guyang greenstone belt, respectively (Jian et al., 2012). As the $2.55 - 2.50$ Ga rocks are exposed across the whole Yinshan Block, the 2.7 Ga magmatic event must have occurred over the whole block, forming an early Neoarchean juvenile crust that experienced partial melting or reworking to form the $2.55 - 2.50$ Ga TTG suites and ultramafic—mafic volcanics. Also as given in Table 3.1, the ~ 2.7 Ga trondhjemite in the Xiwulanbulang area has positive $\varepsilon Hf_{(t)}$ values up to $+8.3$, with most zircon Hf model ages close to its rock-forming age, which suggests that the ~ 2.7 Ga event was a crustal accretion (mantle extraction) event, not a crust-reworking event. As the major components of the $2.55 - 2.50$ Ga rocks are in both the Guyang granite-greenstone belt and Wuchuan high-grade complex, TTGs are generally considered to have been derived from the partial melting of a thickened mafic crust, which implies that the ~ 2.7 Ga juvenile crust should be a mafic-dominant crust.

In summary, the ~ 2.7 Ga magmatic event in the Yinshan Block was a major crustal accretionary period, forming a thick mafic-dominant crust beneath the whole block, whereas the $2.55 - 2.50$ Ga tectonomagmatic event was mainly involved in reworking or partial melting of the ~ 2.7 Ga mafic crust, leading to the formation of numerous $2.55 - 2.50$ Ga TTG rocks and minor $2.6 - 2.5$ Ga volcanic rocks that were subsequently metamorphosed and deformed to form the Guyang granite-greenstone belt and the Wuchuan high-grade complex.

3.4.2 Tectonic Settings

As the ~ 2.7 Ga magmatic event in the Western Block has not been recognized until this study, controversy has not arisen surrounding its tectonic setting. Although the ~ 2.7 Ga mafic—ultramafic rocks have not been found in the block, it can be inferred that this magmatic event must have formed a thick mafic crust that underwent partial melting at $2.55 - 2.50$ Ga to produce voluminous TTG rocks in the Guyang granite-greenstone belt and Wuchuan high-grade complex. One of the most possible tectonic settings for such a large-scale mafic crust-making event was a mantle plume, though it still remains unknown whether such a mantle plume event occurred beneath a continent (underplating) to form the lower crust or within an ocean environment to form an oceanic plateau. As pre-Neoarchean basement rocks have not been found in the Yinshan Block, it is less likely that the Yinshan Block had evolved into a large matured continent with a mafic lower crust and a felsic upper crust by ~ 2.7 Ga. Otherwise, the ~ 2.7 Ga mantle plume event would have caused widespread partial melting of such pre-Neoarchean mafic and felsic crusts to form numerous a sodium-rich TTG suites and potassium-rich granites, respectively, though it cannot completely preclude the possibility that a small matured continent (Yinshan Block) had been completely buried or assimilated by the continental flood basalts of the ~ 2.7 Ga mantle plume event. One of alternative interpretations is that the ~ 2.7 Ga mafic crust-making (mantle plume) event in the Yinshan Block occurred within an oceanic setting, which is consistent with the absence of numerous ~ 2.7 Ga potassium-rich granites in the block. No matter whether the ~ 2.7 Ga mantle plume event in the Yinshan Block occurred beneath a continental or within an ocean, it was a major crust-accretion event forming a mafic crust that then experienced extensive partial melting at $2.55 - 2.50$ Ga to form the $2.55 - 2.50$ Ga TTG rocks and mafic–felsic volcanic rocks in the Guyang granite-greenstone belt and the Wuchuan high-grade complex.

There is a hot debate on the tectonic setting of the $2.55 - 2.50$ Ga tectonomagmatic event in the Yinshan Block. At the center of the controversy is whether the $2.55 - 2.50$ Ga TTG rocks and mafic–felsic volcanic rocks in the Guyang granite-greenstone belt and the Wuchuan high-grade complex formed under the regimes of plate tectonics or mantle plumes (Zhao et al., 1999a,b, 2001a; Jian et al., 2005; Zhang et al., 2006a,b; Chen, 2007; Dong, 2012; Dong et al., 2012a,b; Ma et al., 2013a,b,c,d). The plate tectonics model has recently become more and more popular following recognition of minor $2.55 - 2.52$ Ga sanukitoids (high-Mg diorites) in the Guyang–Wuchuan area (Figure 3.2; Jian et al., 2005; Zhang et al., 2006a,b; Dong, 2012; Ma et al., 2013a). Generally, sanukitoids are considered to have formed in subduction zones and interpreted as the products of partial melting of a mantle wedge which has been previously metasomatised by adakitic melts derived from the melting of a hot, young, subducting slab (Smithies, 2000; Martin et al., 2005). However, sanukitoids in the Yinshan Block are only locally exposed at its southern margin

(Figure 3.2), making up a very small portion of Neoarchean basement rocks in the block; and more importantly, it still remains unknown whether or not these sanukitoids have a genetic connection with voluminous TTG rocks in the Guyang granite-greenstone belt and the Wuchuan high-grade complex. As pointed out by Smithies and Champion (2000), "sanukitoids are typically emplaced in late- to post-kinematic settings sometimes in association with felsic alkaline magmatism, and are either unaccompanied by, or postdate, TTG magmatism, which comprises a much greater proportion of Archean felsic crust." Therefore, it is inappropriate to use sanukitoids to infer the tectonic setting of Neoarchean TTG rocks and mafic–felsic volcanics (greenstones) in the Yinshan Block. In addition, a plate tectonics model has difficulties in explaining the following geological features of the Neoarchean Yinshan Block:

1. Like those in the Eastern Block, TTG rocks in the Yinshan Block do not show any systematic age progression across the block (Zhang et al., 2003; Zhang, 2004; Jian et al., 2005, 2012; Chen, 2007; Ma et al., 2013a–d; Dong, 2012; Dong et al., 2012a), which is inconsistent with migrating or successively accreted magmatic arc models.
2. The 2.55–2.50 Ga volcanic rocks in both the Guyang granite-greenstone belt and Wuchuan high-grade complex are dominated by mafic and felsic components, which contrasts with volcanic assemblages in Phanerozoic continental margin arcs where andesites are dominant (Hamilton, 1998).
3. As mentioned earlier, the Guyang granite-greenstone belt in the Yinshan Block contains komatiites or komatiitic rocks, and the modern-style plate tectonics cannot well explain where and how MgO-rich komatiitic melts with eruption temperatures as high as 1650° are generated in a continental margin arc system.
4. As discussed above, the metamorphic evolution of both the Guyang granite-greenstone belt and Wuchuan high-grade complex is characterized by anticlockwise $P-T$ paths involving IBC. Although the root of a magmatic arc may experience metamorphism with IBC-type anticlockwise $P-T$ paths as discussed by Bohlen (1991), such a magmatic arc should be paired with a subducted slab that undergoes relatively high-pressure metamorphism characterized by clockwise $P-T$ paths involving isothermal decompression, forming the paired metamorphic belts (Brown, 2006, 2007, 2008). This is not the

case in the Yinshan Block where the ~2.5 Ga metamorphism is exclusively characterized by IBC-type anticlockwise $P-T$ paths, without any clockwise $P-T$ paths.

Considering these dilemmas with plate tectonics, Zhao et al. (1999a,b, 2001a) favored a mantle plume model for the formation and evolution of the 2.55 − 2.50 Ga rocks in both the Western and Eastern blocks, with those justifications that have been discussed in Chapter 2 and are not repeated here.

3.4.3 Other Models for the Yinshan Block

Some researchers propose that the Yinshan Block was not a discrete Archean block but was part of a single Archean craton (NCC) that included all Archean lithotectonic units in the Yinshan Block and the Eastern Block (Zhai, 2004, 2011; Zhai et al., 2000, 2010; Zhai and Santosh, 2011; Geng et al., 2010, 2012; Dong, 2012; Jian et al., 2012). Such models were established on the basis of remarked similarities in Neoarchean lithologies and geological events between the Yinshan Block and the Eastern Block. For example, both the Yinshan and Eastern blocks underwent the early Neoarchean crustal growth at ~2.7 Ga followed by widespread crustal reworking in the period 2.55 − 2.50 Ga, though the Yinshan Block does not have any Eoarchean, Paleoarchean, and Mesoarchean records that are preserved in the Eastern Block (see Chapter 2). It is difficult to preclude the possibility that the Yinshan and Eastern blocks existed as a single continent during Archean time, but it is more difficult to prove such a possibility because, if we make such a speculation, we can also speculate that the Yinshan Block and all other cratonic blocks (e.g., South India) that underwent the ~2.7 Ga crustal growth followed by the 2.55 − 2.50 Ga reworking existed as a single continent. As discussed earlier, lithological differences and similarities are not the key to distinguishing different cratonic blocks; the key to identification of different cratonic blocks is recognition of continent−continent collisional belts between terranes. As long as a continent−continent collisional orogen is recognized between two terranes, these two terranes must have belonged to two different continental blocks within a certain period no matter whether or not they have similar lithologies or have undergone similar geological evolution. As the Yinshan (Western) Block is separated from the Eastern Block by the Trans-North China Orogen, a typical continent−continent collisional belt that developed in the

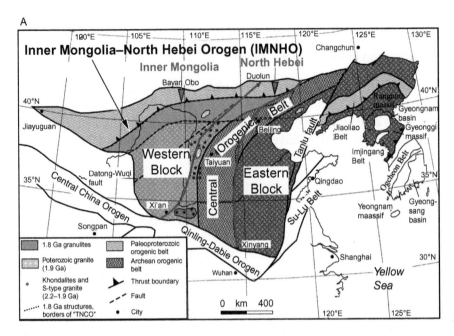

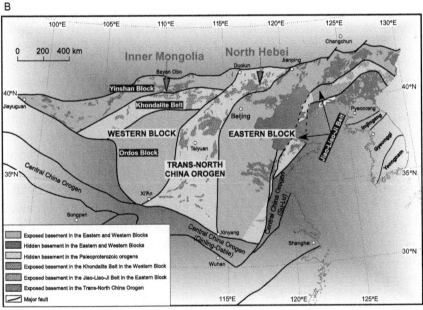

Figure 3.6 The Inner Mongolian portion of the IMNHO of Kusky et al. (2007) shown in (A) occupies the same spatial position as the Yinshan Block of Zhao et al. (2005) shown in (B).

period 2.56–1.85 Ga (see Chapter 4), the Yinshan (Western) and Eastern blocks should be regarded as two independent cratonic blocks at least in the period 2.56–1.85 Ga.

Kusky and Li (2003) regarded the Yinshan Block as part of a Paleoproterozoic orogen named the "Inner Mongolia–North Hebei Orogen (IMNHO)", which they assumed to have formed by collision between the northern margin of the NCC and an exotic arc terrane at ~2.3 Ga, and later collided with the Columbia (Nuna) supercontinent at 1.92–1.85 Ga (Kusky et al., 2007; Kusky and Santosh, 2009). According to Kusky and Li (2003) and Kusky et al. (2007), the Paleoproterozoic   IMNHO   consists   of   granitoid   gneisses (2.47–2.42 Ga), gabbro–diorite complexes, minor ultramafic rocks, and voluminous supracrustal sequences (2.64–2.30 Ga). However, such Paleoproterozoic rock assemblages only crop out in the north Hebei segment of the proposed IMNHO, represented by the Hongqingyingzi and Dantazi "groups or complexes" (Geng et al., 1997, 1999; Wang et al., 2002), but are not exposed in the Inner Mongolia segment of the IMNHO. As shown in Figure 3.6A and B, the Inner Mongolia segment of the IMNHO occupies the same spatial position as the Yinshan Block. As discussed above, the Yinshan Block is a typical Neoarchean continental block and thus should not be regarded as a segment of the so-called "Paleoproterozoic IMNHO". Therefore, the Neoarchean nature of the Yinshan Block does not support the Kusky and Li (2003) model that the northern margin of the NCC collides with an exotic arc terrane at 2.3 Ga and then collides with the Columbia (Nuna) supercontinent at 1.92–1.85 Ga. Even for the north Hebei portion of the IMNHO, there are no convincing data suggesting that it underwent an arc-continent collision at ~2.3 Ga, though it experienced a regional metamorphic event at 1.9–1.8 Ga (Mao et al., 1999). Zhao et al. (2012) interpreted this metamorphic event to be part of the collision between the Eastern and Western blocks at ~1.85 Ga, as the north Hebei segment of the proposed IMNHO coincides with the northernmost part of the Trans-North China Orogen (Figure 3.6A and B). Therefore, the so-called "Paleoproterozoic IMNHO" on the northern margin of the NCC may not have existed (Zhao et al., 2012).

CHAPTER 4

Paleoproterozoic Amalgamation of the North China Craton

4.1 INTRODUCTION

Available data show that the late Paleoproterozoic (2.1−1.8 Ga) was the first period in Earth's history in which large-scale continent−continent collision under the regime of modern-style plate tectonics occurred, forming the Himalayan-type collisional orogens on nearly every craton, including the 2.1−2.0 Ga Transamazonian orogen in South America, 2.1−2.0 Ga Eburnean orogen in West Africa, 1.95−1.85 Ga Trans-Hudson Orogen or its equivalents (Taltson−Thelon, Wopmay, New Quebec, Foxe, Makkovik, Ungava, and Torngat Orogens) in North America, 1.9−1.8 Ga Nagssugtoqidian Orogen in Greenland, 1.9−1.8 Ga Kola−Karelia, Volhyn-Central Russian and Pachelma Orogens in Baltica (including East Europe), 1.9−1.8 Ga Akitkan Orogen in Siberia, ∼2.0 Ga Limpopo Belt in South Africa, ∼2.0 Ga Capricorn Belt in Western Australia, etc. These orogens are considered to have recorded global-scale collisional events that led to the assembly of a Paleo-Mesoproterozoic supercontinent, named "Columbia" (Rogers and Santosh, 2002; Zhao et al., 2002a) or "Nuna" (Hoffman, 1997).

Extensive field-based structural, metamorphic, geochemical, geo-chronological, and geophysical investigations carried out in the last decade have demonstrated that like most other cratonic blocks, the North China Craton (NCC) was also formed by amalgamation of a number of microcontinental blocks in the late Paleoproterozoic (1.95−1.85 Ga), which was coincident with the global 2.1−1.8 Ga col-lisional events that led to the assembly of the proposed Paleo-Mesoproterozoic Columbia (Nuna) supercontinent. A major advance-ment in understanding the Paleoproterozoic amalgamation of the NCC was made following recognition of three major Paleoproterozoic tectonic belts, namely, the Khondalite Belt, Jiao-Liao-Ji Belt, and Trans-North China Orogen (TNCO) in the western, eastern, and cen-tral parts of the craton, respectively (Figure 1.4; Zhao et al., 2005).

Precambrian Evolution of the North China Craton. DOI: http://dx.doi.org/10.1016/B978-0-12-407227-5.00004-3

This chapter will summarize available data demonstrating that the Khondalite Belt and TNCO are typical Himalayan-type continent—continent collisional belts that formed during the Paleoproterozoic amalgamation of microcontinental blocks to form the NCC, and the Jiao-Liao-Ji Belt is a rifting-and-collision belt within the Eastern Block which underwent rifting to form an incipient oceanic basin that was closed upon itself through subduction and collision at ~1.9 Ga.

4.2 KHONDALITE BELT IN THE WESTERN BLOCK

4.2.1 Major Lithologies of the Khondalite Belt

The Paleoproterozoic Khondalite Belt is a ~1000 km long east-west-trending belt that extends from the Helanshan and Qianlishan complexes in the west, through the Daqingshan and Wulashan complexes in the middle, to the Jining Complex in the east (Figure 4.1). The belt is dominated by high-grade Al-rich gneiss (pelitic granulite), garnet-bearing quartzite, felsic paragneiss, calc-silicate rock and marble, which are collectively called the "khondalite series" in the Chinese literature. It has long been considered that the sedimentary protoliths of the khondalite series in the Khondalite Belt were deposited in the Neoarchean (Qian and Li, 1999; Li et al., 1999) or even Mesoarchean

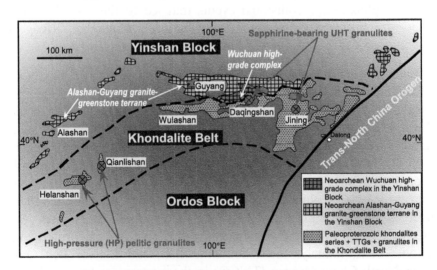

Figure 4.1 A map showing the distribution of the Khondalite Belt which divides the Western Block into the Yinshan Block in the north and the Ordos Block in the south. The map also shows the localities of the high-pressure (HP) pelitic granulites and ultrahigh temperature (UHT) rocks in the belt. After Zhao et al. (2005).

(Yang et al., 2000) and were metamorphosed ~2.5 Ga (Lu et al., 1992, 1996; Qian and Li, 1999; Li et al., 1999). However, recent sensitive high-resolution ion microprobe (SHRIMP) and laser-ablation induced coupled plasma-mass spectrometry (LA-ICP-MS) zircon dating data indicate that they were deposited and metamorphosed in the Paleoproterozoic, with detrital zircon ages mostly ranging from 2.3 to 2.0 Ga and metamorphic zircon ages of 1.95–1.87 Ga (Table 4.1; Wu et al., 2006; Xia et al., 2006a,b, 2008; Wan et al., 2006a; Dong et al., 2007; Yin et al., 2009, 2011; Dong, 2012; Ma et al., 2012). The sedimentary protoliths of the "khondalite series" are generally considered to have been deposited on a stable continental margin (Condie et al., 1992; Lu and Jin, 1993; Zhao et al., 1999a, 2005), though some researchers argue that they were sourced from a ca. 2.18–2.00 Ga continental arc based on their positive $\varepsilon Hf(t)$ values (Dan et al., 2012). Associated with the khondalite series rocks are large volumes of S-type granites that formed at the postorogenic or anorogenic stage (Jin et al., 1991; Lu and Jin, 1993; Guo et al., 1999; Zhao et al., 1999a; Yin et al., 2009, 2011; Zhong, 2010). In the Daqingshan and Wulashan complexes, in addition to the khondalite series that is dominated by garnet–sillimanite gneisses or granulites, there are large amounts of garnet–biotite paragneisses are closely associated with end-Archean tonalite–trondhjemite–granodiorite (TTG) gneisses and mafic granulites, which are considered to represent the exposure of the Wuchuan high-grade complex in the Khondalite Belt (Dong, 2012). Also associated with the garnet–biotite paragneisses, end-Archean TTG gneisses and mafic granulites in the Daqingshan and Wulashan complexes are Paleoproterozoic adakites, sanukitoids, and Closepet granites, dated at 2435 ± 12 Ma, 2426 ± 41 Ma and 2416 ± 8 Ma, respectively (Zhong, 2010), which are considered to have formed on a northward subduction zone beneath the southern margin of the Yinshan Block (Zhao et al., 2005). The sedimentary protoliths of the khondalite series may not have been deposited on the southern margin of the Neoarchean Yinshan Block as the khondalite series rocks are dominated by Paleoproterozoic detrital zircons with ages ranging between 2.3 and 2.0 Ga, with few Neoarchean detrital zircons (Table 4.1; Xia et al., 2006a,b; Wan et al., 2006a; Yin et al., 2009, 2011; Zhou and Geng, 2009), different from the garnet–biotite gneisses that contain a considerable amount of Neoarchean detrital zircons (Ma et al., 2012; Dong et al., 2013). As the khondalite series rocks mainly crop out marginal to the Ordos Basin (Xia et al., 2009), it is reasonable to infer that they

Table 4.1 Summary of Zircon U−Pb Ages of Khondalites from the Khondalite Belt, NCC

Location	Sample No.	Rock Type	Detrital Zircons	Metamorphic Zircons	Method	Reference
Helanshan Complex	HL1-2	Garnet-bearing quartzite	2020 Ma	1958 Ma	LA-ICP-MS	Yin et al. (2011)
	HL2-3	Garnet−biotite gneiss	2020 Ma	1953 Ma 1869 Ma	LA-ICP-MS	Yin et al. (2011)
	HL2-4	Garnet−sillimanite−biotite gneiss	2116−2006 Ma	1952 Ma 1865 Ma	LA-ICP-MS	Yin et al. (2011)
	HL2-5	Sillimanite−garnet−cordierite gneiss	2855−2019 Ma	1955 Ma 1865 Ma	LA-ICP-MS	Yin et al. (2011)
	HL3-1	Garnet−cordierite gneissic migamatite	2034 Ma	1946 Ma	LA-ICP-MS	Yin et al. (2011)
	HL3-2	Sillimanite−garnet−cordierite gneiss	2196−2003 Ma	1963 Ma	LA-ICP-MS	Yin et al. (2011)
	HL3-3	Garnet−cordierite gneiss	2151−2002 Ma 2530 Ma (1 spot)	1962 Ma	LA-ICP-MS	Yin et al. (2011)
	HL0702-2	Garnet−cordierite gneiss	2039 Ma 2949 and 2520 Ma (two spots)		SHRIMP	Zhou and Geng (2009)
	HL0706-1	Garnet−cordierite gneiss	2171−2040 Ma	1950 Ma	SHRIMP	Zhou and Geng (2009)
	HL0707-10	Garnet−biotite−plagioclase gneiss	2144−2017 Ma		SHRIMP	Zhou and Geng (2009)
Qianlishan Complex	QL85-2	Garnet−sillimanite−biotite gneiss	2035−1999 Ma	1952 Ma 1921 Ma	LA-ICP-MS	Yin et al. (2009)
	QL87-1	Garnet−sillimanite−cordierite gneiss	2082−2021 Ma	1950 Ma	LA-ICP-MS	Yin et al. (2009)
	QL91-3	Garnet−feldspar−quartz gneiss	2035 Ma	1954 Ma	LA-ICP-MS	Yin et al. (2009)

Complex	Sample	Rock type			Method	Reference
Wulashan–Daqingshan Complex	QL93-6	Garnet–sillimanite–biotite gneiss	2253–2028 Ma	1941 Ma	LA-ICP-MS	Yin et al. (2009)
	QL95-9	Garnet–feldspar quartzite	2028 Ma	1955 Ma	LA-ICP-MS	Yin et al. (2009)
	QL86-1	Garnet–sillimanite leucoleptite	2275–2010 Ma	1953 Ma	LA-ICP-MS	Yin et al. (2009)
	WL007	Metapelitic gneiss	2363–2070 Ma / 2562 Ma (one spot)	1965 Ma / 1875 Ma	LA-ICP-MS	Xia et al. (2006b)
	WL011	Sillimanite–garnet–biotite gneiss	2097–2068 Ma	1927 Ma / 1880 Ma	LA-ICP-MS	Xia et al. (2006b)
	WL016	Meta-quartzite	2008 Ma		LA-ICP-MS	Xia et al. (2006b)
	WL020	Garnet-bearing metapelitic gneiss	2200–2050 Ma / 2502 Ma (one spot)	1954 Ma	LA-ICP-MS	Xia et al. (2006b)
Jining Complex	SDT0101	Sillimanite–biotite gneiss		1838 Ma	SHRIMP	Wan et al. (2006a)
	SDT0104	SIllimanite–biotite gneiss	2019 Ma	1873 Ma	SHRIMP	Wan et al. (2006a)
	JS0102	SIllimanite–garnet gneiss	2330 Ma	1861 Ma	SHRIMP	Wan et al. (2006a)
	01M020	Sillimanite–garnet gneiss		1902 Ma	LA-ICP-MS	Xia et al. (2006a)
	01M038	Sillimanite–garnet gneiss	2175–2031 Ma	1957 Ma / 1887 Ma	LA-ICP-MS	Xia et al. (2006a)
	01M041	Sillimanite–garnet gneiss		1945 Ma	LA-ICP-MS	Xia et al. (2006a)
	01M053	Cordierite–sillimanite gneiss	2099 Ma	1842 Ma	LA-ICP-MS	Xia et al. (2006a)

represent stable continental margin deposits surrounding the Ordos Block. In contrast, the Paleoproterozoic garnet—biotite gneisses that contain large volumes of Neoarchean detrital zircons were most likely to have deposited on the southern margin of the Archean Yinshan Block. Therefore, the khondalite series rocks are in tectonic contact with late Neoarchean TTG gneisses and mafic granulites and Paleoproterozoic adakites, sanukitoids, and Closepet granites in the Khondalite Belt.

4.2.2 Metamorphic Evolution of the Khondalite Belt

Most of the khondalite series rocks from the Khondalite Belt are medium-pressure pelitic granulites that preserve four distinct mineral assemblages (M1—M4). M1 is represented by inclusions of plagioclase + biotite + quartz ± kyanite ± rutile within the M2 garnet porphyroblasts; M2 represents the growth of garnet porphyroblasts and matrix plagioclase + biotite + quartz + sillimanite ± ilmenite; M3 is represented by cordierite coronas and cordierite + sillimanite or cordierite + spinel symplectites surrounding garnet porphyroblasts; and M4 represents retrograde minerals biotite + chlorite replacing garnet, K-feldspar + sericite + chlorite replacing cordierite or andalusite + muscovite dissecting the main foliation (Jin, 1989; Jin et al., 1991; Lu, 1991; Lu and Jin, 1993; Jin and Li, 1994, 1996; Liu, 1996, 1997; Liu and Shen, 1999). These mineral assemblages and their thermobarometric estimates define clockwise $P-T$ paths involving near-isothermal decompression (Figure 4.2A), reflecting a continental collisional setting (Zhao et al., 1999a). This forms the justification for the Zhao et al. (2002b, 2005) model that interprets the Khondalite Belt as a Paleoproterozoic continent—continent collisional belt along which the Yinshan Block in the north collided with the Ordos Block in the south to form the Western Block. This model has been further supported by recent discoveries of high-pressure pelitic granulites in the Qianlishan and Helanshan complexes in the western sector of the Khondalite Belt (Figure 4.1; Zhou et al., 2010; Yin, 2010; Yin et al., 2014a,b). High-pressure pelitic granulites, characterized by a key peak assemblage of kyanite-K-feldspar-garnet (O'Brien and Rötzler, 2003), have been considered as a hallmark of continental subduction and collision tectonics because only tectonic processes related to subduction and continent—continent collision can transport the sedimentary precursors of the pelitic granulites to a lower crustal depth where they experience high-pressure granulite-facies metamorphism. This is consistent with

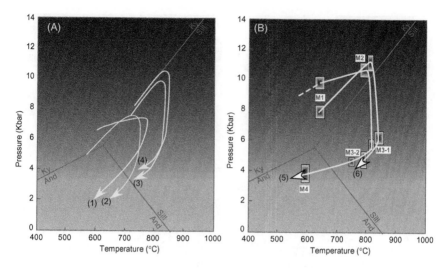

Figure 4.2 Metamorphic P–T paths of (A) medium-pressure and (B) high-pressure pelitic granulites in the Khondalite Belt: (1) medium-pressure pelitic granulites from the Helanshan Complex (Zhao et al., 1999a); (2) medium-pressure pelitic granulites from the Daqingshan–Wulashan Complex (Jin et al., 1991); (3) medium-pressure pelitic granulites from the Datong Complex (Liu et al., 1997a); (4) medium-pressure pelitic granulites from the Jining Complex (Wang et al., 2011a,b); (5) high-pressure pelitic granulites from the Helanshan Complex (Yin et al., 2014a); and (6) high-pressure pelitic granulites from the Qianlishan Complex (Yin et al., 2014b).

the clockwise *P−T* paths involving isothermal decompression reconstructed for the high-pressure pelitic granulites from both the Helanshan and Qianlishan complexes (Yin et al., 2014a,b; Figure 4.2B).

In the Helanshan Complex, the high-pressure pelitic granulites contain the early prograde (M1), peak high-pressure (M2), post-peak decompression (M3), and later retrograde (M4) assemblages, of which the prograde assemblage (M1) is plagioclase + biotite + quartz + muscovite + kyanite + ilmenite, preserved as mineral inclusions within the core of garnet (Figure 4.3A and B). The peak high-pressure granulite-facies assemblage (M2) is kyanite + K-feldspar + garnet + plagioclase + quartz + biotite + ilmenite in the matrix (Figure 4.3C). The post-peak (M3) was involved in two stages of decompression of which the first is represented by the formation of the cordierite + sillimanite symplectites (M3-1) followed by the development of the cordierite ± spinel coronas (M3-2), both replacing garnet porphyroblasts (Figure 4.3D−F). The later retrograde metamorphism (M4) resulted in the formation of euhedral staurolite crystals in association with chlorite in the matrix (Figure 4.3G). Pseudosection

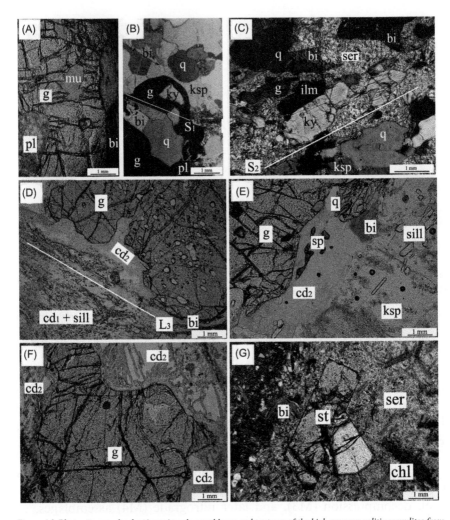

Figure 4.3 Photomicrographs showing mineral assemblages and textures of the high-pressure pelitic granulites from the Helanshan Complex (Yin et al., 2014a). (A) Muscovite, biotite, and plagioclase inclusions within the core of a porphyroblastic garnet (plane-polarized light). (B) Kyanite, biotite, and quartz inclusions within garnet (cross-polarized light). (C) Kyanite, K-feldspar, ilmenite, biotite, garnet, and quartz in the matrix (cross-polarized light). (D) A cordierite corona between a garnet grain and the cordierite + sillimanite symplectite (plane-polarized light). (E) Relatively large prismatic sillimanite crystals in the matrix and a cordierite + spinel corona surrounding garnet porphyroblasts (plane-polarized light). (F) A cordierite corona surrounding garnet porphyroblasts (plane-polarized light). (G) Late retrograde staurolite and chlorite in the matrix (cross-polarized light). Abbreviations: mu = muscovite; g = garnet; bi = biotite; q = quartz; kip = K-feldspar; ilm = ilmenite; ser = sericite; ky = kyanite; sill = sillimanite; pl = plagioclase; cd = cordiorite; st = staurolite; chl = chlorite; S1 = foliation forming during the first phase of deformation; S2 = foliation forming during the first phase of deformation; L3 = lineation forming during the third phase of deformation.

modeling in the $Na_2O-CaO-K_2O-FeO-MgO-SiO_2-H_2O-TiO_2-O$ (NCKFMASHTO) system constrains the $P-T$ conditions of the M1, M2, M3-1, M3-2, and M4 assemblages at 9.3−9.7 kbar/

656−674°C,   10.2−11.2 kbar/792−805°C,   5.5−5.7 kbar/810−820°C, 4.5−5.0 kbar/780−785°C, and 3.4−4.4 kbar/580−610°C, respectively (Yin et al., 2014b). These mineral assemblages and their $P-T$ conditions define a clockwise $P-T$ path involving isothermal decompression (ITD) and isobaric cooling (Figure 4.4).

A similar $P-T$ path was established for the high-pressure pelitic granulites from the Qianlishan Complex, which underwent the prograde (M1), peak (M2), and post-peak decompression (M3) stages. The M1 assemblage is quartz + plagioclase + muscovite + biotite +

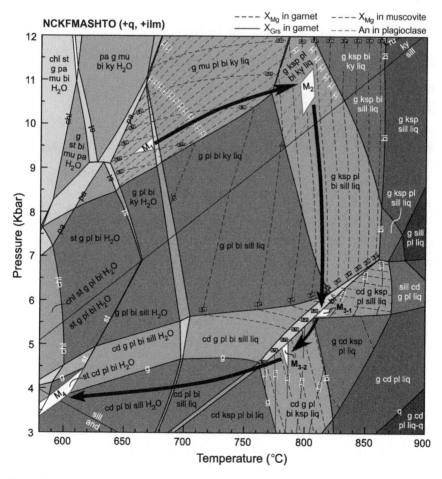

Figure 4.4 P−T *path of the high-pressure pelitic granulites from the Helanshan Complex reconstructed using pseudosection modeling based on the bulk-rock composition:* $H_2O = 5.409$, $SiO_2 = 60.389$, $Al_2O_3 = 13.653$, $CaO = 1.142$, $MgO = 6.487$, $FeO = 8.651$, $K_2O = 2.650$, $Na_2O = 0.884$, $TiO_2 = 0.517$, $O = 0.216$ in mol.% (Yin et al., 2014a).

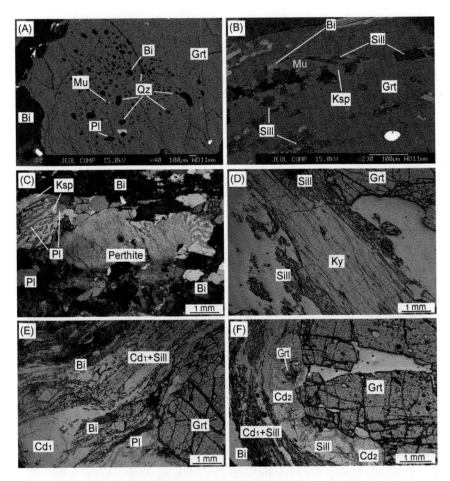

Figure 4.5 Representative microstructural features of high-pressure pelitic granulites from the Qianlishan complex (Yin et al., 2014b). (A) Mineral inclusions of biotite, plagioclase, muscovite, and quartz within the core of a garnet porphyroblast (sample S82-1; BSE image). (B) Some sillimanite inclusions and a small muscovite flake within K-feldspar enclosed in garnet (BSE image). (C) K-feldspar containing thin lamellae of plagioclase as perthite coexisting with biotite and plagioclase in the matrix (cross-polarized light). (D) Small sillimanite crystals replacing large tabular kyanite in contact with garnet (plane-polarized light). (E) The cordierite + sillimanite symplectite in the matrix containing minor relict biotite flakes, suggesting that the formation of the cordierite + sillimanite symplectite at the expenses of biotite (plane-polarized light). (F) The cordierite + sillimanite symplectite in the matrix and cordierite corona surrounding garnet porphyroblasts (plane-polarized light).

sillimanite, which occur as mineral inclusions within the core of garnet porphyroblasts (Figure 4.5A and B). The peak stage (M2) is represented by the mantle growth of garnet porphyroblasts and matrix minerals kyanite, K-feldspar, biotite, plagioclase, and quartz (Figure 4.5C). The decompression stage (M3) is indicated by sillimanite replacing kyanite (Figure 4.5D), the formation of the cordierite + sillimanite symplectite in the matrix (M3-1; Figure 4.5E), and the cordierite corona replacing garnet (M3-2; Figure 4.5F). These mineral assemblages and their $P-T$ estimates based on the assemblage stability fields of $P-T$ pseudosection constructed in NCKFMASHTO define a clockwise $P-T$ path involving near-isothermal decompression (Figure 4.6).

The ITD-type clockwise $P-T$ paths of the high-pressure pelitic granulites from the Helanshan and Qianlishan complexes are in accord with those $P-T$ paths reconstructed for the medium-pressure pelitic granulites/gneisses in the Khondalite Belt (Figure 4.2A and B; see Zhao et al., 1999a and references wherein), suggesting that the whole Khondalite Belt experienced tectonic processes that are characterized by initial

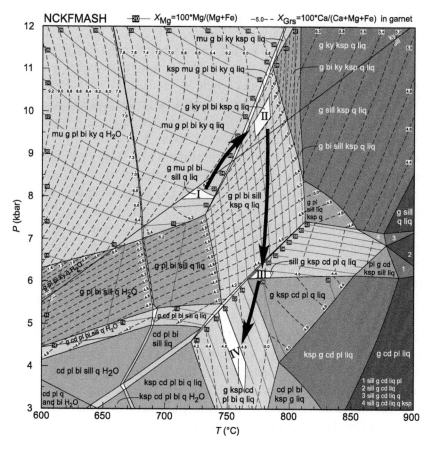

Figure 4.6 P$-$T *path of the high-pressure pelitic granulites from the Qianlishan Complex reconstructed by pseudosection modeling based on the bulk-rock composition:* $H_2O = 5.75$, $SiO_2 = 60.64$, $Al_2O_3 = 13.72$, $CaO = 1.15$, $MgO = 6.51$, $FeO = 8.69$, $K_2O = 2.66$, $Na_2O = 0.89$ *in mol.% (Yin et al., 2014b).*

burial and crustal thickening (M1−M2), followed by isothermal exhumation (M3) and final retrogression and cooling (M4). This strongly supports the Zhao et al. (2002b, 2005) model that the Khondalite Belt was a continent−continent collisional belt along which the Yinshan and Ordos blocks amalgamated to form the Western Block.

4.2.3 Timing of Collision Between the Yinshan and Ordos Blocks

To determine the timing of collision between the Yinshan and Ordos blocks to form the Western Block, a number of researchers have applied the SHRIMP and LA-ICP-MS techniques to date metamorphic zircons from both the high-pressure and medium-pressure pelitic granulites from the Khondalite Belt, and the results show that the metamorphism of these pelitic granulites occurred at ∼1.95 Ga (Table 4.1; Dong et al., 2007, 2013; Yin et al., 2009, 2011; Zhou and Geng, 2009; Zhao et al., 2010a; Li et al., 2011a). For example, metamorphic zircons from four high-/medium-pressure pelitic granulite samples collected from the Qianlishan Complex yielded consistent weighted mean $^{207}Pb/^{206}Pb$ ages of 1949 ± 12, 1954 ± 22, 1941 ± 24, and 1955 ± 21 Ma (Table 4.1; Yin et al., 2009). S-type granites from the Qianlishan Complex yielded zircon ages of ∼1875 Ma (Yin et al., 2009), interpreted as the time of late exhumation of the Khondalite Belt. Similarly, metamorphic zircons from seven high-/medium-pressure pelitic granulite samples collected from the Helanshan Complex yielded weighted mean $^{207}Pb/^{206}Pb$ ages of 1962 ± 14, 1963 ± 15, 1958 ± 7, 1953 ± 7, 1955 ± 15, 1952 ± 9, and 1946 ± 13 Ma (Table 4.1; Yin et al., 2011), similar to those metamorphic ages obtained for the Qianlishan Complex (Yin et al., 2009), though minor metamorphic zircons gave ages of ∼1.86 Ga, (Yin et al., 2011). Dong et al. (2007) obtained similar SHRIMP U−Pb metamorphic ages for the Helanshan Complex. For the Wulashan and Daqingshan complexes in the middle sector of the Khondalite Belt (Figure 4.1), Dong et al. (2013) obtained two populations of metamorphic ages with the earlier set at ∼1.95 Ga and younger ages at ∼1.86 Ga, similar to the metamorphic ages of the Qianlishan and Helanshan complexes. For the Jining Complex, Li et al. (2011a) also recognized metamorphic ages of both ∼1.95 and ∼1.86 Ga for the non-ultrahigh temperature (UHT) pelitic granulites. These data demonstrate that the collision between the Yinshan and Ordos blocks along the Khondalite Belt occurred at ∼1.95 Ga, earlier than the collision of the Western and Eastern blocks along the TNCO at ∼1.85 Ga. This was supported by a study on metamorphic zircons from the Huai'an and Jining complexes (Zhao et al., 2010a; Li et al., 2011a), which are located in the conjunction area of the Khondalite Belt with the TNCO (Figure 1.4) and thus should have undergone two metamorphic events. As shown in Figure 4.7A, some zircons from these complexes have a dark (low luminescent) core and double overgrowth rims, and both rims are structureless, highly luminescent and have low Th/U ratios (<0.01), typical of a metamorphic origin. The inner rims yield a weighted mean $^{207}Pb/^{206}Pb$ age of 1946 ± 26 Ga (Figure 4.7B), whereas the outer rims gave a weighted mean $^{207}Pb/^{206}Pb$ age of 1850 ± 15 Ga (Figure 4.7C). The age of the inner rims is interpreted as the approximate timing of the collision between the Yinshan and Ordos blocks to form the Khondalite Belt, whereas the age of the outer rims is the time of collision between the Western and Eastern blocks to form the TNCO (Zhao et al., 2010a).

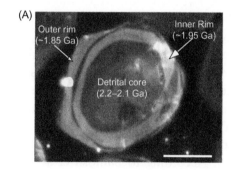

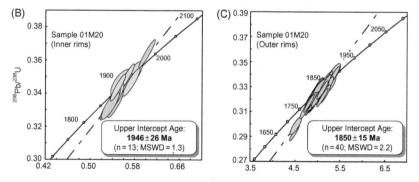

Figure 4.7 (A) Cathodoluminescence (CL) image showing that some zircons from the Jining Complex, which is located in the conjunction area of the Khondalite Belt with the TNCO, have a dark detrital core and double meta-morphic overgrowth rims (Zhao et al., 2010a). (B) Concordia diagrams of SHRIMP results for the inner meta-morphic zircon rims shown in (A). (C) Concordia diagrams of SHRIMP results for the outer metamorphic zircon rims shown in (A).

4.2.4 UHT Metamorphism in the Khondalite Belt

In addition to high- and medium-pressure pelitic granulites, the Khondalite Belt also contains minor sapphirine-bearing UHT pelitic granulites (Figure 4.1; Guo et al., 2006, 2012; Santosh et al., 2006, 2007a, b, 2008, 2012; Liu et al., 2008b, 2010; Zhao, 2009; Santosh, 2010; Jiao and Guo, 2011; Jiao et al., 2011, 2013a,b), which were first discovered by Jin (1989) in the Daqingshan area and were further investigated by Liu et al. (2000a), but at that time they did not well locate the *in situ* outcrops of these rocks and did not recognize them as UHT rocks. Guo et al. (2006) are the first to find the *in situ* sapphirine−sillimanite−spinel-bearing UHT granulites at Dongpo in the Daqingshan Complex, and in the same year, Santosh et al. (2006) reported sapphirine−orthopyroxene−sillimanite UHT rocks at Tuguiwula in the Jining Complex, which show distinct differences in chemical composition, mineral assemblage, and metamorphic $P-T$ conditions from the Dongpo UHT rocks in the Daqingshan Complex. The Dongpo UHT rocks are extremely unsaturated in SiO_2 (<39%wt), with low MgO and high Al_2O_3, and the representative assemblage is sapphirine + sillimanite + garnet + spinel + plagioclase + biotite + cordierite + rutile/ilmenite, without quartz and orthopyroxene, whereas the Tuguiwula UHT rocks are saturated in SiO_2 and high in MgO, with a typical UHT assemblage of sapphirine + orthopyroxene + sillimanite + garnet + spinel + plagioclase + biotite + quartz + cordierite. Moreover, the maximum temperature of the Tuguiwula UHT rocks was estimated at 1030°C (Santosh et al., 2007b), much higher than that of the Dongpo UHT rocks (910−980°C; Guo et al., 2012). According to Santosh et al. (2006, 2007a, 2009a,b), the Tuguiwula UHT metamorphism underwent three stages (M1, M2, and M3), of which M1 occurred at 1030°C, forming the orthopyroxene + sapphirine + sillimanite + garnet + quartz + biotite + spinel assemblage, M2 occurred under $P-T$ conditions of $P > 10$ kbar; $T > 950°C$, forming the orthopyroxene + sillimanite + garnet + quartz assemblage, and M3 took place at $P < 10$ kbar and $T \approx \sim 950°C$, forming orthopyroxene + cordierite + garnet + quartz. These mineral assemblages and their $P-T$ conditions define an anticlockwise $P-T$ path (Figure 4.8), reflecting the initial cooling of an ultrahot metamorphic terrain followed by rapid exhumation, but the whole terrain was still at the middle to lower crustal level. For the Dongpo UHT rocks, Guo et al. (2012) recognized five metamorphic assemblages (M0, M1, M2, M3, and M4), of which M0 is garnet (core) +

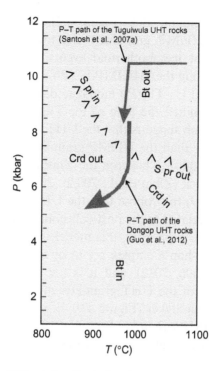

Figure 4.8 P–T *paths of the UHT rocks from Tuguiwula in the Jining area (Santosh et al., 2007a) and Dongpo in the Daqingshan area (Guo et al., 2006, 2012).*

sillimanite + spinel + biotite + plagioclase ± rutile/ilmenite, preserved as inclusion-type minerals, which did not result from the UHT metamorphism, but formed before the UHT metamorphism. M1 is garnet (mantle/rim) + matrix   minerals   sapphirine + spinel + sillimanite + biotite + plagioclase. M2 and M3 are sapphirine + plagioclase and spinel + plagioclase symplectites, respectively, both of which resulted from the post-peak decompression; and M4 represents the retrogressive metamorphism, forming retrograde biotite (Guo et al., 2012). Using the THERMOCALC pseudosection modeling technique, Guo et al. (2012) and Jiao et al. (2013a) defined clockwise P−T paths involving isothermal decompression, which just makes a supplementary P−T segment that was not recorded by the Tuguiwula UHT rocks (Figure 4.8). Combined the two P−T paths together, the whole metamorphic evolution of the UHT rocks in the Khondalite Belt is characterized by anticlockwise P−T paths (Figure 4.8), different from the ITD-type clockwise P−T paths reconstructed for the high-/medium-pressure pelitic granulites in the Khondalite Belt (Figure 4.2), implying that the UHT metamorphism and HP/MP granulite-facies metamorphism may have resulted from different tectonothermal events, which is supported by metamorphic ages. Using the SHRIMP U−Pb zircon and electron probe microanalysis (EPMA) U−Th−Pb monazite dating techniques, Santosh et al. (2007a,b) determined the time of the Tuguiwula UHT metamorphism at 1.92 Ga, which suggests that the UHT metamorphism occurred about 20−30 Ma later than the HP/MP granulite-facies metamorphism, which resulted from collision between the Yinshan and Ordos blocks to form the Western Block (Zhao et al., 2002b, 2005). This led Guo et al. (2006) and Zhao (2009) to propose that the UHT metamorphism of the Khondalite Belt was related to the post-collisional mantle upwelling. One line of robust evidence supporting this model is that most UHT granulites from the Daqingshan Complex are in direct contact with gabbroic dykes (Guo et al., 2006, 2012), and it is the same case in the Jining Complex where most of the UHT granulites are spatially in direct contact with meta-gabbroic dykes (Figure 4.9), of which one meta-gabbroic dyke yielded a SHRIMP zircon U−Pb age of 1927 ± 12 Ma (G.C. Zhao, unpublished data), nearly at the same time as the timing of the UHT

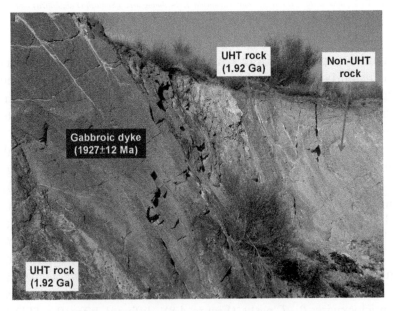

Figure 4.9 A field photograph showing the close relationship between the 1.92 Ga UHT rocks and ~1.92 Ga gabbroic dyke (Zhao and Zhai, 2013).

metamorphism (Santosh et al., 2007b). In addition to the post-collisional mantle upwelling model (Guo et al., 2006; Zhao, 2009; Zhao et al., 2012; Jiao et al., 2013b), some other models have also been proposed, including ridge subduction (Santosh et al., 2008; Peng et al., 2010, 2011, 2012a), mantle plume bombarding carbonated tectosphere (Santosh et al., 2008, 2009a,b), extensive exhumation of the lower crust (Zhai, 2009), and paired metamorphic belt (Liu and Li, 2009), which have been well summarized by Santosh et al. (2012). Of these models, the ridge subduction model has recently become more and more popular (Santosh et al., 2008; Peng et al., 2010, 2011, 2012a; Guo et al., 2012). A problem with the ridge subduction model is that the UHT metamorphism should have happened earlier than the final collision of the Yinshan and Ordos blocks, but as discussed above, the UHT metamorphism occurred some 30 − 20 Ma later than the high-/medium-pressure granulite-facies metamorphism that resulted from the Yinshan−Ordos collision. This led Peng et al. (2010, 2011) to propose that the UHT metamorphism in the Khondalite Belt resulted from the westward ridge subduction of an old ocean between the Eastern and Western blocks. Peng et al. (2012a) proposed that the ridge subduction also accounted for the emplacement of the 1950−1880 Ma gabbro-norite−charnockite−S-type granite suite and the eruption of similar-aged volcanic rocks in the Jining area. However, such a ridge subduction model cannot well explain the UHT metamorphism in the Daqingshan Complex, which is far away from the eastern margin of the Western Block (Figure 4.1).

4.3 JIAO-LIAO-JI BELT IN THE EASTERN BLOCK

The Jiao-Liao-Ji Belt divides Eastern Block into the Longgang (Yanliao) Block in the northwest and the Langrim Block in the southeast (Figure 4.10; Zhao et al., 2005). The belt consists of granitic and mafic intrusions and sedimentary and volcanic successions metamorphosed from greenschist to lower amphibolite facies. The sedimentary and volcanic successions, including the Fenzishan and Jingshang groups in the Jiaobei massif, the North and South Liaohe groups in the eastern Liaoning Peninsula, the Ji'an and Laoling groups in southern Jilin, and the Macheonayeong Group in North Korea (Figure 4.10), are transitional from a basal clastic-rich and lower bimodal-volcanic sequence, through a middle carbonate-rich sequence,

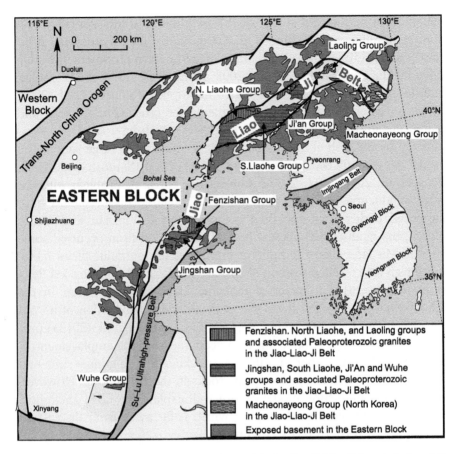

Figure 4.10 Geological map showing the spatial distribution of the Jiao-Liao-Ji Belt, which can be further subdivided into the northern and southern zones; the former comprises the Laoling, North Liaohe, and Fenzishan groups, whereas the latter consists of the Ji'an, South Liaohe, and Jingshan groups. After Zhao et al. (2005).

to an upper pelite-rich sequence (Luo et al., 2004, 2006, 2008; Li et al., 2004a,b, 2005, 2006, 2011b, 2012; Lu et al., 2006, 2008b; Li and Zhao, 2007; Zhou et al., 2008; Tam et al., 2011, 2012a–c). The Paleoproterozoic Wuhe Group in Anhui Province has lithologies and ages similar to those of the Fengzishan and Jingshan groups in Eastern Shandong and thus is considered as the southwestern extension of the Jiao-Liao-Ji Belt (Guo and Li, 2009). Tectonically associated with the sedimentary and volcanic units in the Jiao-Liao-Ji Belt are numerous Paleoproterozoic mafic and granitoid intrusions, of which the former consists of gabbros and dolerites metamorphosed from greenschist facies to amphibolite facies (Li et al., 2005), whereas the latter includes deformed A-type granites and undeformed alkaline syenites and rapakivi granites (Hao et al., 2004; Li et al., 2004a,b, 2005, 2006; Lu et al., 2006; Li and Zhao, 2007). Recent SHRIMP and LA-ICP-MS U−Pb zircon ages show that pretectonic (gneissic) granites in the Jiao-Liao-Ji Belt were emplaced in the period ca. 2.2−2.0 Ga, forming a source for the sedimentary successions that were deposited within a narrow period between 2.0 and 1.95 Ga (Luo et al., 2004, 2008; Lu et al., 2006; Li and Zhao, 2007; Zhou et al., 2008; Tam et al., 2011). Available data also indicate that the Jiao-Liao-Ji Belt underwent two metamorphic events, with the earlier one occurred at 1.93−1.90 Ga, whereas the latter one took place at ∼1.87 Ga (Luo et al., 2004, 2008; Lu et al., 2006; Li and Zhao, 2007; Zhou et al., 2008; Tam et al., 2011). Stratigraphically, the North Liaohe Group in eastern Liaoning is well correlated with the Laoling Group in southern Jilin and the Fenzishan in eastern Shandong, whereas the South Liaohe Group in eastern Liaoning can also be stratigraphically correlated with the Ji'an Group in southern Jilin and the Jingshan Group in eastern Shandong (Zhang and Yang, 1988; He and Ye, 1998; Zhao et al., 2005). Thus, the Jiao-Liao-Ji Belt is further divisible into the southern and northern zones, of which the former consists of the Jingshan, South Liaohe, and Ji'an groups, whereas the latter comprises the Fenzishan, North Liaohe, and Laoling groups (Figure 4.10; He and Ye, 1998; Zhao et al., 2005). The two zones, separated by a series of ductile shear zones and faults (Li et al., 2005), show contrasting metamorphic $P−T$ paths: anticlockwise $P−T$ paths for the southern zone and clockwise $P−T$ paths for the northern zone (Figure 4.11; Lu et al., 1996; He and Ye, 1998; Li et al., 2001). However, these $P−T$ paths were established based on $P−T$ conditions estimated using traditional geothermobarometers and thus their reliability needs further test.

A number of competing tectonic models have been proposed for the formation and evolution of the Paleoproterozoic Jiao-Liao-Ji Belt, from those invoking arc-continent collision (Bai, 1993; Faure et al., 2004; Lu et al., 2006) to those involving the opening and closing of an intracontinental rift (Zhang and Yang, 1988; Peng and Palmer, 1995; Li et al., 2004a,b, 2005, 2006, 2011b, 2012; Luo et al., 2004, 2008; Li and Zhao, 2007). On the basis of structural, metamorphic, geochemical, and geochronological studies on the North and South Liaohe groups and associated granitoids, Bai (1993) was the first to propose that the Jiao-Liao-Ji Belt was an arc-continent collisional belt along

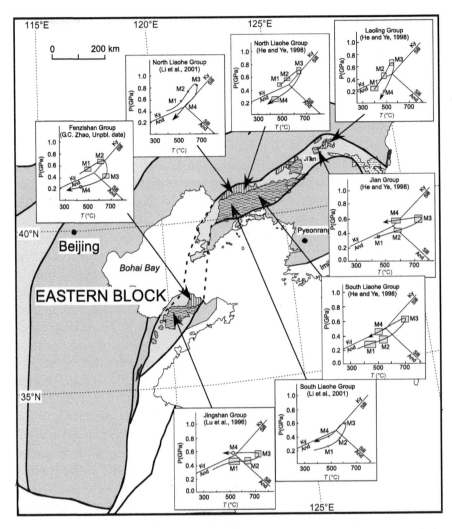

Figure 4.11 Metamorphic P–T paths of the Fenzishan, Jingshan, North Liaohe, South Liaohe, Laoling and Ji'an groups in the Paleoproterozoic Jiao-Liao-Ji belt (Zhao et al., 2011a and references wherein).

which an eastern island arc (Langrim Block) and a western Archean block (Longgang Block) amalgamated to form a coherent cratonic block in the Paleoproterozoic. In this model, the North and South Liaohe groups were considered to have developed within a N-S-trending back-arc basin, which was closed to form the Jiao-Liao-Ji Belt during arc-continent collision in the Paleoproterozoic. However, this arc-continent collision model cannot well explain the absence of calc-alkaline igneous associations characterizing modern magmatic arcs in the Jiao-Liao-Ji Belt. Faure et al. (2004) made a further modification on this arc-continent collision model by interpreting mafic—ultramafic rocks in the North Liaohe Group as an active continental arc above a southward subduction zone that developed between a northern Archean block (Longgang Block) and a southern block that is largely composed of the South Liaohe Group. Later, this magmatic arc belt was overthrust upon the northern Archean basement during the arc-continent collision. This model implies that the North and South Liaohe groups did not develop simultaneously but belonged to two discrete terranes. This is not supported by recent zircon ages and Hf isotope data that indicate that two groups developed coevally on a single Archean continental block (Luo et al., 2008).

Zhang and Yang (1988) were the first to have applied the rift closure model interpret the formation of the Jiao-Liao-Ji Belt. In this model, the Longgang and Langrim blocks were considered as a single Archean block that underwent a rifting event in the early Paleoproterozoic, leading to the formation of a rift basin in which sedimentary-volcanic rocks formed and granitoid and mafic intrusions were emplaced. This rift basin was closed in the late Paleoproterozoic, leading to deformation and metamorphism of the sedimentary-volcanic rocks in the basin, forming the Jiao-Liao-Ji Belt. This model has been advocated by Sanzhong Li and his collaborators (Li et al., 2004a,b, 2005, 2006, 2011b, 2012; Hao et al., 2004; Luo et al., 2004, 2006, 2008; Li and Zhao, 2007) because they think that the model is supported by the following lithotectonic elements in the Jiao-Liao-Ji Belt: (i) voluminous A-type granites; (ii) bimodal volcanics, represented by large volumes of meta-mafic volcanics (greenschists and amphibolites) and meta-rhyolites (Zhang and Yang, 1988; Peng and Palmer, 1995); (iii) borate deposits of a nonmarine origin, which have many similarities to those borate-bearing successions in other Proterozoic rifting environments, for example, the Upper Proterozoic Damaran Orogen of South Africa (Peng and Palmer, 1995); (iv) similar-aged Neoarchean TTG gneisses and mafic dyke swarms on the opposite sides of the belt (Zhang and Yang, 1988); and (v) low-pressure-type, anticlockwise, $P-T$ paths of the Ji'an, South Liaohe, and Jingshan groups in the southern zone of the belt (Lu et al., 1996; He and Ye, 1998), which are not consistent with a continent—continent collision model. However, the rift closure model has difficulties in explaining the polyphase compressive deformation and clockwise $P-T$ paths of the Laoling, North Liaohe, and Fenzishan groups in the northern zone of the Jiao-Liao-Ji Belt.

Recently, Paleoproterozoic high-pressure pelitic granulites have been discovered in the Jingshan Group of the Jiao-Liao-Ji Belt. As mentioned earlier, the presence of high-pressure pelitic granulites is of particular significance to recognition of subduction and collisional orogens because only plate tectonic processes involving subduction and continent–continent collision can bring the sedimentary precursors of pelitic granulites down to a lower crustal depth where they experience high-pressure granulite-facies metamorphism. Thus, the presence of high-pressure pelitic granulites in the Jiao-Liao-Ji Belt implies that the formation and evolution of the belt must have been involved in sub-duction- or collision-related tectonic processes (Bai, 1993; Faure et al., 2004). Therefore, even though the Jiao-Liao-Ji Belt was initialized from a rift basin (Zhang and Yang, 1988; Luo et al., 2004, 2008; Li et al., 2004a, 2005, 2006; Li and Zhao, 2007), this rift basin must have developed into an ocean basin at least in its southern segment, where the oceanic lithosphere was subducted, leading to the final closure of the ocean basin with the formation of the high-pressure pelitic granu-lites. In this sense, the Jiao-Liao-Ji Belt can be regarded as a Paleoproterozoic rift-and-collision belt in the Eastern Block, which underwent an extensional and rifting event in the period 2.2–1.9 Ga, leading to the opening of an incipient ocean that broke up the Eastern Block into the Longgang Block in the northwest and the Langrim Block in southeast, which were reassembled along the Jiao-Liao-Ji Belt through subduction and collision (Zhou et al., 2008; Tam et al., 2011, 2012a–c; Zhao and Zhai, 2013; Zhao et al., 2012).

4.4 TRANS-NORTH CHINA OROGEN: FINAL AMALGAMATION OF THE WESTERN AND EASTERN BLOCKS

This nearly S-N-trending, ∼1200 km long and 100–300 km wide, belt was first recognized by Zhao et al. (1998) who initially called it the "Central Zone," and later named it the "Trans-North China Orogen (TNCO)" (Figure 4.12; Zhao et al., 2001a). Li et al. (2000) and Kusky and Li (2003) modified the boundaries of the TNCO and called it the "Central Orogenic Belt" (Figures 1.6 and 3.6). The belt consists of upper amphibolite to granulite-facies gneiss complexes, including the Taihua (TH), Fuping (FP), Hengshan (HS), Huai'an (HA), Xuanhua (XH), and Chengde (CD) complexes, and greenschist to lower amphibolite-facies granite-greenstone terranes including the Zhongtiao (ZT), Dengfeng (DF), Zanhuang (ZH), Wutai (WT), and North Hebei

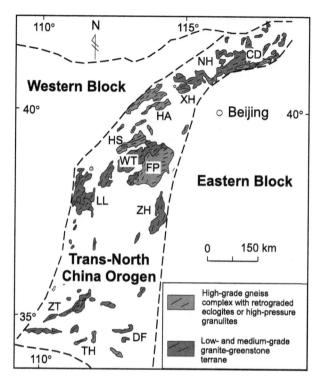

Figure 4.12 Simplified tectonic map showing the distribution of metamorphic complexes in the TNCO. Abbreviations for metamorphic complexes: CD, Chengde; DF, Dengfeng; FP, Fuping; HA, Huai'an; HS, Hengshan; LL, Lüliang; NH, Northern Hebei; TH, Taihua; WT, Wutai; XH, Xuanhua; ZH, Zanhuang; ZT, Zhongtiao. After Zhao et al. (2000a).

(NH) complexes (Figure 4.12). Available isotopic data show that the major metamorphism and deformation of the TNCO occurred in the Paleoproterozoic, but most lithologies in the orogen formed in the late Neoarchean and early Paleoproterozoic (25.0–24.5 Ma), whereas the Paleoproterozoic lithologies in the TNCO are mainly exposed in its middle-southern segment, which was named the "Jinyu Belt" (Figure 1.5) by Zhai and Peng (2007).

4.4.1 Major Lithologies of the TNCO

Table 4.2 lists the major lithologies of the TNCO and their ages, which include the following major lithotectonic elements: (i) minor 2.8–2.7 Ga grey gneisses and xenocrystic zircons that are interpreted as the remnants of the preorogenic old continental basement, represented by the 2.8–2.7 Ga Dengfeng grey gneisses; (ii) 2550–2520 Ma granitoid plutons in the low- to medium-grade granite-greenstone

Table 4.2 Summary of Isotopic Ages for Major Lithotectonic Elements in the TNCO

Sample	Description	Age (Ma)	Method	Sources
Ages of Remnants of Old Continental Crust				
Xuanhua Complex				
JN0737	Pyroxenite xenolith in the Hannuoba basalt	2715 ± 7	LA-ICP-MS	Jiang et al. (2010)
Huai'an Complex				
Z9167	Xenocrystic zircons from a grey gneiss (Wagoutai)	2761, 2656	SHRIMP	Liu et al. (1997b)
Hengshan Complex				
990843	Grey trondhjemitic gneiss (Changchenggou)	2712 ± 2	SHRIMP	Kröner et al. (2005b)
	Same sample	2711.1 ± 0.3	SGE	Kröner et al. (2005b)
990838	Well foliated grey biotite gneiss (Dashiyu)	2701 ± 5.5	SHRIMP	Kröner et al. (2005b)
980811	Grey granodioritic gneiss (Changchenggou)	2697.1 ± 0.3	SGE	Kröner et al. (2005b)
980824	Finely layered, fine-grained biotite gneiss (Dashiyu)	2670.6 ± 0.4	SGE	Kröner et al. (2005b)
FP52	Migmatiteic granites (leucosome)	2686 ± 7	LA-ICP-MS	Faure et al. (2007)
Wutai Complex				
95-PC-114	Xenocrystic zircons from a meta-andesite (Ekou)	2679 ± 16	SHRIMP	Wilde et al. (2004a)
95-PC-94	Xenocrystic zircons from the Lanzhisan granite	2702 ± 14	SHRIMP	Wilde et al. (1997)
Fuping Complex				
95-PC-65	Xenocrystic zircons from the Longquanguan granite	2763 ± 10	SHRIMP	Wilde (2002)
	Same sample	2660 ± 7	SHRIMP	Wilde (2002)
FP50	Xenocrystic zircons from a biotite gneiss (Diebuan)	2708 ± 8	SHRIMP	Guan et al. (2002)
FP260	Oldest detrital zircons from the Wanzi supracrstals	2826 ± 8	SHRIMP	Zhao et al. (2002c)
	Same sample	2686 ± 6	SHRIMP	Zhao et al. (2002c)
08CWW05	Xenocrystic zircons from 2514 Ma grey gneiss	2753 ± 35	ICP-MS	Han et al. (2012)
Dengfeng Complex				
D001	A zircon core from 2512 ± 12 Ma meta-dacite	2945 ± 44	SHRIMP	Kröner et al. (1988)
	Xenocrystic zircons from the same sample	2650–2945	SHRIMP	Kröner et al. (1988)

(Continued)

Table 4.2 (Continued)

Sample	Description	Age (Ma)	Method	Sources
Zhongtiao Complex				
Unknown	A xenocrystic zircon from Jiangxian Group	2770 ± 16	SHRIMP	Sun et al. (1993b)
Taihua Complex				
LS0417-2	Tonalitic gneiss	2829 ± 13	SHRIMP	Liu et al. (2009a)
LS0417-4	Tonalitic gneiss	2832 ± 11	SHRIMP	Liu et al. (2009a)
LS0417-1	Amphibolite	2838 ± 35	SHRIMP	Liu et al. (2009a)
LS0417-1	Amphibolite	2845 ± 23	SHRIMP	Liu et al. (2009a)
TW006-1	Two xenocrystic zircons from pelitic gneiss (Washi)	2726 ± 15	SHRIMP	Wan et al. (2006b)
Lu003	Two zircons from a tonalitic gneiss (Lushan)	2841 ± 6	SGE	Kröner et al. (1988)
	One zircon from the same sample	2806 ± 7	SGE	Kröner et al. (1988)
Ages of Late Neoarchean Granitoids from Granite-Greenstone Terranes (2560–2520 Ma)				
Wutai Granite-Greenstone Terrane				
95-PC-34	Deformed, pink, medium-grained granitoid (Ekou)	2566 ± 13	SHRIMP	Wilde et al. (1997)
95-19	Deformed, pink, medium-grained granitoid (Ekou)	2555 ± 6	SHRIMP	Wilde et al. (1997)
WC7	Foliated, coarse-grained tonalite (Chechang-Beitai)	2552 ± 11	SHRIMP	Wilde et al. (2005)
95-PC-6B	Foliated, coarse-grained tonalite (Chechang-Beitai)	2551 ± 5	SHRIMP	Wilde et al. (2005)
WC6	Fine-grained granodiorite (Chechang-Beitai)	2546 ± 6	SHRIMP	Wilde et al. (2005)
WC5	Coarse-grained granodiorite (Chechang-Beitai)	2538 ± 6	SHRIMP	Wilde et al. (2005)
95-PC-94	Deformed, coarse-grained granitoid (Lanzhishan)	553 ± 8	SHRIMP	Wilde et al. (1997)
95-PC-96	Deformed, coarse-grained granitoid (Lanzhishan)	2537 ± 10	SHRIMP	Wilde et al. (1997)
95-PC-98	Deformed medium-grained monzogranite (Shifo)	2531 ± 4	SHRIMP	Wilde et al. (2005)
95-PC-76	Deformed, medium-grained granitoid (Guangmishi)	2531 ± 5	SHRIMP	Wilde et al. (2005)
95-PC-62	Grey, medium-grained granodiorite (Wangjiahui)	2520 ± 9	SHRIMP	Wilde et al. (2005)
95-PC-63	Grey, medium-grained granodiorite (Wangjiahui)	2517 ± 12	SHRIMP	Wilde et al. (2005)

(Continued)

Table 4.2 (Continued)

Sample	Description	Age (Ma)	Method	Sources
WL12	Sheared porphyritic granitoid (Longquanguan)	2543 ± 7	SHRIMP	Wilde et al. (1997)
WN11	Sheared tonaltic granitoid (Yushuwan Village)	2541 ± 14	SHRIMP	Wilde et al. (1997)
WL9	Sheared porphyritic granitoid (Longquanguan)	2540 ± 18	SHRIMP	Wilde et al. (1997)
Zhongtiao Granite-Greenstone Terrane				
ZT4020-1	Foliated trondhjemite (Sushui)	2553 ± 14	SHRIMP	Tian et al. (2005)
Dengfeng Granite-Greenstone Terrane				
DF07−8	Trondhjemitic gneiss	2521 ± 21	LA-ICP-MS	Diwu et al. (2011)
DF07−10	Trondhjemitic gneiss	2510 ± 20	LA-ICP-MS	Diwu et al. (2011)
DF07−15	Tonalitic gneiss	2587 ± 55	LA-ICP-MS	Diwu et al. (2011)
DF07−24	Tonalitic gneiss	2544 ± 10	LA-ICP-MS	Diwu et al. (2011)
09DF−19	Dioritic gneiss	2529 ± 14	LA-ICP-MS	Diwu et al. (2011)
Ages of Late Nearchean Greenstone-Type Volcanics (2530−2515 Ma)				
Wutai Granite-Greenstone Terrane				
95-PC-114	Meta-andesite, Zhuangwang "Formation" (Ekou)	2529 ± 10	SHRIMP	Wilde et al. (2004a)
96PC-119	Meta-andesite, Zhuangwang "Formation" (Ekou)	2513 ± 8	SHRIMP	Wilde et al. (2004a)
95-PC-115	Meta-andesite, Baizhiyan "Formation" (Ekou)	2524 ± 10	SHRIMP	Wilde et al. (2004a)
WT13	Meta-rhyolite, Hongmenyan "Formation" (S-T)	2533 ± 8	SHRIMP	Wilde et al. (2004a)
WT17	Meta-rhyodacite Hongmenyan "Formation" (S-T)	2524 ± 8	SHRIMP	Wilde et al. (2004a)
WT9	Meta-dacite, Hongmenyan "Formation" (S-T)	2523 ± 9	SHRIMP	Wilde et al. (2004a)
WT12	Meta-rhyodacite, Hongmenyan "Formation" (S-T)	2516 ± 10	SHRIMP	Wilde et al. (2004a)
95-PC-55c	Meta-rhyolite, Gaofan "Subgroup" (Xiazhuang)	2528 ± 6	SHRIMP	Wilde et al. (2004a)
Dengfeng Granite-Greenstone Terrane				
D001	15 zircons from a meta-rhyodacite (Dengfeng)	2512 ± 12	SHRIMP	Kröner et al. (1988)
09DF−18	Magmatic zircons from amphibolites	2547 ± 14	LA-ICP-MS	Diwu et al. (2011)

(Continued)

Table 4.2 (Continued)

Sample	Description	Age (Ma)	Method	Sources
Ages of Late Neoarchean to Paleoproterozic High-Grade TTG Gneisses (2520–2480 Ma)				
Shangyi (Northern Hebei) TTG Gneisses				
Z9116	Tonalitic gneiss	2512 ± 19	SHRIMP	Wang et al. (2009a)
Xuanhua TTG Gneisses				
Z9116	Granodioritic gneiss (Zhangquanzhuang)	2515 ± 7	SHRIMP	Liu et al. (1997b)
Z9121	Granodioritic gneiss (Guzuizi Village)	2503 ± 23	SGD	Liu et al. (1997b)
XW03	Tonalitic gneiss	2499 ± 21	SHRIMP	Liu et al. (2012a)
JN0703	Tonalitic gneiss	2486 ± 20	LA-ICP-MS	Jiang et al. (2010)
Huai'an TTG Gneisses				
Z9167	Eight zircons from a grey tonaltic gneiss (Wagoutai)	2526 ± 13	SHRIMP	Geng et al. (1997)
08MQG30	Tonalitic gneiss	2503 ± 17	LA-ICP-MS	Wang et al. (2010a)
M21	Tonalitic gneiss	2515 ± 20	SHRIMP	Zhao et al. (2008a)
M19	Trondhjemitic gneiss	2499 ± 19	SHRIMP	Zhao et al. (2008a)
M23	Granodioritic gneiss	2440 ± 26	SHRIMP	Zhao et al. (2008a)
05LF75	Tonalitc gneiss	2503 ± 14	LA-ICP-MS	Liu et al. (2009b)
05LF86	Tonalitc gneiss	2471 ± 44	LA-ICP-MS	Liu et al. (2009b)
05LF96	Trondhjemitic gneiss	2506 ± 13	LA-ICP-MS	Liu et al. (2009b)
05LF96	Trondhjemitic gneiss	2542 ± 6	CAMECA	Liu et al. (2012a)
05LF38	Dioritic gneiss	2494 ± 41	LA-ICP-MS	Liu et al. (2009b)
05LF77	Dioritic gneiss	2507 ± 17	LA-ICP-MS	Liu et al. (2009b)
05LF77	Dioritic gneiss	2518 ± 8	CAMECA	Liu et al. (2012a)
05LF51	Dioritic gneiss	2508 ± 14	CAMECA	Liu et al. (2012a)
Hengshan TTG Gneisses				
980814	Dioritic gneiss (Changchenggou)	2479 ± 3	SHRIMP	Kröner et al. (2005b)
	Same sample	2478.2 ± 0.3	SGE	Kröner et al. (2005b)

(Continued)

Table 4.2 (Continued)

Sample	Description	Age (Ma)	Method	Sources
990803	Dioritic gneiss with melt patches (Dashiyu)	2475 ± 2	SHRIMP	Kröner et al. (2005b)
990859	Dioritic gneiss (Xiaoshiyu)	2506 ± 5	SHRIMP	Kröner et al. (2005b)
	Same sample	2504.6 ± 0.3	SGE	Kröner et al. (2005b)
980802	Homogeneous tonalitic gneiss (Dashiyu)	2500.5 ± 0.3	SGE	Kröner et al. (2005b)
990871	Tonalitic gneiss, small side valley of Ruyuegou	2502.3 ± 0.3	SGE	Kröner et al. (2005b)
HG5	Garnetiferous trondhjemitic gneiss (Yanmenguan)	2520 ± 10	SHRIMP	Kröner et al. (2005b)
HG6	Trondhjemitic gneiss, roadcut near Yanmenguan	2526 ± 12	SHRIMP	Kröner et al. (2005b)
HG7	Trondhjemitic gneiss, a roadcut near Yanmenguan	2507 ± 4	SHRIMP	Kröner et al. (2005b)
990845	Foliated trondhjemitic gneiss (Xiaoshiyu)	2504.4 ± 0.4	SGE	Kröner et al. (2005b)
980838	Trondhjemitic gneiss (Ruyuegou)	2503.0 ± 0.3	SGE	Kröner et al. (2005b)
990847	Grey, unveined trondhjemitic gneiss (Yanmenguan)	2524 ± 8	SHRIMP	Kröner et al. (2005b)
	Same sample	2521.7 ± 0.4	SGE	Kröner et al. (2005b)
990854	Trondhjemitic gneiss (Xiaoshiyu)	2504.6 ± 0.3	SGE	Kröner et al. (2005b)
980809	Foliated pegmatitic red granite gneiss (Dashiyu)	2501 ± 3	SHRIMP	Kröner et al. (2005b)
	Same sample	2503.5 ± 0.3	SGE	Kröner et al. (2005b)
980825	Fine-grained, strained granite-gneiss (Xiaoshiyu)	2496.3 ± 0.3	SGE	Kröner et al. (2005b)
980833	Granitic gneiss (Xiaoshiyu)	2492.4 ± 0.3	SGE	Kröner et al. (2005b)
990873	Foliated granite gneiss from migmatite (Guaner)	2499 ± 6	SHRIMP	Kröner et al. (2005b)
	Same sample	2497.6	SGE	Kröner et al. (2005b)
980803	Homogeneous felsic gneiss (Dashiyu)	2502.3 ± 0.6	SGE	Kröner et al. (2005b)
990821	Finely layered, fine-grained biotite gneiss (Dashiyu)	2526 ± 04.7	SHRIMP	Kröner et al. (2005b)
96PC153	Homogeneous tonalitic gneiss, near Yixingzhai	2513 ± 15	SHRIMP	Wilde (2002)
96PC154	Quartz dioritic gneiss, a roadcut near Yixingzhai	2499 ± 4	SHRIMP	Wilde (2002)

(Continued)

Table 4.2 (Continued)				
Sample	Description	Age (Ma)	Method	Sources
09LYK03	Tonalitic gneiss	2505 ± 15	LA-ICP-MS	Zhao et al. (2010b)
09LYK05	Tonalitic gneiss	2511 ± 15	LA-ICP-MS	Zhao et al. (2010b)
Fuping TTG Gneisses				
FG1	Layered tonalitic gneiss, a roadcut near Xicaokou	2523 ± 14	SHRIMP	Zhao et al. (2002c)
FP50	Foliated, hornblende-rich tonalitic gneiss (Diebuan)	2520 ± 20	SHRIMP	Guan et al. (2002)
FP54	Trondhjemitic gneiss collected from Pingyang	2513 ± 12	SHRIMP	Guan et al. (2002)
FP217	Trondhjemitic gneiss collected from Tuanpokou	2499 ± 9.5	SHRIMP	Zhao et al. (2002c)
FP216	Granodioritic gneiss collected from Tuanpokou	2486 ± 8	SHRIMP	Zhao et al. (2002c)
FP08	Granodioritic gneiss collected from Xicaokou	2475 ± 8	SHRIMP	Guan et al. (2002)
FP236	Mylonitized monzongranitic gneiss (near Ciyu)	2510 ± 22	SHRIMP	Zhao et al. (2002c)
FP224	Mylonitized pegmatitic dyke (Xinzhuang)	2507 ± 11	SHRIMP	Zhao et al. (2002c)
FP133	Migmatitic granite	2481 ± 7	LA-ICP-MS	Trap et al. (2009b)
FP135	Migmatitic granite	2456 ± 11	LA-ICP-MS	Trap et al. (2008)
Yunzhongshan TTG Gneisses				
M132	Tonalitic gneiss	2499 ± 9	SHRIMP	Zhao et al. (2008b)
FP354	Migmatite	2535 ± 41	LA-ICP-MA	Trap et al. (2009b)
Ages of Paleoproterozoic (2360–2000 Ma) Granitoids				
Huai'an Complex				
M28	Igneous zircons from a granitic gneiss	2036 ± 16	SHRIMP	Zhao et al. (2008a)
Hengshan Complex				
980806	Fine-grained granitic orthogneiss (Dashiyu)	2358.7 ± 0.5	SGE	Kröner et al. (2005b)
990850	Layered trondhjemitic gneiss (Dashiyu)	2329.7 ± 0.6	SGE	Kröner et al. (2005b)
HG4	Pegmatitic granite-gneiss (Dashiyu)	2331 ± 36	SHRIMP	Kröner et al. (2005b)
980881	Pegmatite cutting older gneisses (Xiaoshiyu)	2248.5 ± 0.5	SGE	Kröner et al. (2005b)
980844	Red anatectic granite (Changchenggou)	2113 ± 8	SHRIMP	Kröner et al. (2005b)

(Continued)

Table 4.2 (Continued)

Sample	Description	Age (Ma)	Method	Sources
	Same sample	2112.3 ± 0.6	SGE	Kröner et al. (2005b)
09LYK06	Hornblende monzogranite	2084 ± 4	CAMECA	Zhao et al. (2010b)
09LYK06	Hornblende monzogranite	2052 ± 17	LA-ICP-MS	Zhao et al. (2010b)
09LYK13	Biotite monzogranite	2083 ± 15	CAMECA	Zhao et al. (2010b)
09LYK13	Biotite monzogranite	2060 ± 18	LA-ICP-MS	Zhao et al. (2010b)
Wutai Complex				
D2	Coarse-grained Dawaliang porphyritic granite	2176 ± 12	SHRIMP	Wilde (2002)
95-PC-50	Pink Phase of the Wangjiahui granite	2117 ± 17	SHRIMP	Wilde et al. (2005)
95-PC-51	Pink Phase of the Wangjiahui granite	2116 ± 16	SHRIMP	Wilde et al. (2005)
95-PC-60	Pink Phase of the Wangjiahui granite	2084 ± 16	SHRIMP	Wilde et al. (2005)
HT21	Basaltic andesite from the Hutuo Group	2140 ± 14	SHRIMP	Du et al. (2010b)
Fuping Complex				
FP188-2	Nanying granitoid gneisses, 10 km south of Fuping	2077 ± 13	SHRIMP	Zhao et al. (2002c)
FP204	Nanying granitoid gneisses near Dianfang	2024 ± 21	SHRIMP	Zhao et al. (2002c)
FP30	Nanying granitoid gneisses collected from Gangnan	2045 ± 64	SHRIMP	Guan et al. (2002)
08CWW04	(Nanying) granitoid gneisses	2114 ± 42	ICP-MS	Han et al. (2012)
Zanhuang Complex				
Z19-1	Granite	2109 ± 10	SHRIMP	Yang et al. (2011)
Lüliang Complex				
L9738	Hornblende-bearing granitoid gneiss (Ehutan)	2151 ± 12	SGD	Geng et al. (2000)
L9773	Biotite-bearing granitoid gneiss (Fangshan County)	2152 ± 35	SGD	Geng et al. (2000)
L9735	Monzogranite (Jiaoliushen)	2031 ± 47	SGD	Geng et al. (2000)
M110	Porphyritic granitic gneiss (Gaijiazhuang gneisses)	2375 ± 10	SHRIMP	Zhao et al. (2008b)

(Continued)

Table 4.2 (Continued)

Sample	Description	Age (Ma)	Method	Sources
M112	Tonalitic gneiss (Chijianling-Guandishan gneisses)	2199 ± 11	SHRIMP	Zhao et al. (2008b)
M112	Granodioritic gneiss (Chijianling gneisses)	2180 ± 7	SHRIMP	Zhao et al. (2008b)
M112	Monzogranitic gneiss (Chijianling gneisses)	2173 ± 7	SHRIMP	Zhao et al. (2008b)
FP285	Orthogneiss	2152 ± 36	LA-ICP-MA	Trap et al. (2009b)
05LL008-3	Gneissic tonalite	2180 ± 15	SHRIMP	Liu et al. (2009c)
05LL01-5	Gneissic granodiorite	2180 ± 7	SHRIMP	Liu et al. (2009c)
05LL01-5	Gneissic monzogranite	2090 ± 18	SHRIMP	Liu et al. (2009c)
Zhongtiao Complex				
Unknown	Tonalitic gneiss	2321 ± 2	SGD	Sun et al. (1993b)
Ages of Paleoproterozoic Volcano-Sedimentary Rocks				
Huai'an Complex				
08MQG23	Detrital zircons from a meta-sedimentary rock	2064–2046 (2057 ± 22)	ICP-MS	Wang et al. (2010a)
Hutuo Group in the Wutai Complex				
HTG-10	A felsic tuffaceous rock at Taihuai: an older group	2180 ± 4	SHRIMP	Wilde et al. (2004b)
	A younger population from the same sample	2087 ± 9	SHRIMP	Wilde et al. (2004b)
Wanzi Supracrustal Assemblage in the Fuping Complex				
FP62	A paragneiss collected from Pingyang	2097 ± 46	SHRIMP	Guan et al. (2002)
FP249	The youngest concordant zircon from a paragneiss	2109 ± 5	SHRIMP	Zhao et al. (2002c)
FP260	One young discordant zircon from a paragneiss	2097 ± 6	SHRIMP	Zhao et al. (2002c)
Lüliang Complex				
L9732	Paragneiss from the Jieheku Group (Jiaoliushen)	2028 ± 45	SGD	Geng et al. (2000)
LY1-15	Amphibolite from the Lüliang Group (Jinzhouying)	2051 ± 68	SGD	Yu et al. (1997)
LY1-15	Meta-rhyolite from the Lüliang Group (Dujiagou)	2099 ± 41	SGD	Yu et al. (1997)
L9747	Paragneiss from the Lüliang Group (Dujiagou)	2080 ± 11	SGD	Geng et al. (2000)

(Continued)

Table 4.2 (Continued)

Sample	Description	Age (Ma)	Method	Sources
L9710	Mica-bearing feldspathic quartzite (Yejishan Group)	2124 ± 38	SGD	Geng et al. (2000)
Zhongtiao Complex				
Unknown	Meta-rhyolitic tuff from Jiangxian Group	2115 ± 6	SHRIMP	Sun et al. (1993b)
Unknown	Meta-rhyolitic tuff from Jiangxian Group	2116 ± 1	SGD	Sun et al. (1993b)
Crystallization Ages of Paleoproterozoic Pretectonic Mafic Dykes (Now Amphiolite or Granulites)				
Huai'an Complex				
03XW01	Medium-pressure granulite-facies mafic dyke	1973 ± 4	SHRIMP	Peng (2005)
08MQG26	Ingneous zircons from a HP granulite	1964 ± 38	ICP-MS	Wang et al. (2010a)
M17	Igneous zircons from a HP granulite	1964 ± 60	SHRIMP	Zhao et al. (2008a)
Hengshan Complex				
Ch020901	High-pressure granulite-facies mafic dyke (Dashiyu)	1915 ± 4	SHRIMP	Kröner et al. (2006)
Ch020902	High-pressure granulite-facies mafic dyke (Dashiyu)	1914 ± 2	SHRIMP	Kröner et al. (2006)
WS85	Yanmenguan mafic–ultramafic intrusion	2193 ± 5	SHRIMP	Wang et al. (2010b)
Wutai Complex				
02SX009	Amphibolite-facies mafic dyke (Hengling)	2147 ± 5	SHRIMP	Peng (2005)
Ages of Metamorphism (1880–1800 Ma)				
Chengde Complex				
By98020	Meta-zircons from a HP granulite (Dantazi village)	1817 ± 17	SGD	Mao et al. (1999)
Xuanhua Complex				
XW99-8	Garnet-whole rock for HP granulite (Xiwangshan)	1842 ± 38	Sm–Nd	Guo and Zhai (2001)
WM99-7	Garnet-whole rock for amphibolite (Xiwangshan)	1856 ± 26	Sm–Nd	Guo and Zhai (2001)
XW22	Meta-zircons from HP granulites (Xiwangshan)	1803 ± 9	SHRIMP	Guo et al. (2005)
	Meta-zircons from the same sample	1872 ± 16	SHRIMP	Guo et al. (2005)
XW23	Meta-zircons from HP granulites (Xiwangshan)	1819 ± 16	SHRIMP	Guo et al. (2005)
XW03	Meta-zircons from tonalitic gneiss	1826 ± 9	SHRIMP	Liu et al. (2012a)

(Continued)

Table 4.2 (Continued)

Sample	Description	Age (Ma)	Method	Sources
JN0703	Meta-zircons from mafic granulite	1793 ± 13	LA-ICP-MS	Jiang et al. (2010)
JN0731	Meta-zircons from amphibolites	1814 ± 6	LA-ICP-MS	Jiang et al. (2010)
Huai'an Complex				
MQ91-8	Garnet-whole rock for HP granulite (Manjinggou)	1813 ± 23	Sm–Nd	Guo et al. (2005)
MJ35	Meta-zircons from HP granulites (Manjinggou)	1817 ± 12	SHRIMP	Guo et al. (2005)
MJ36	Meta-zircons from HP granulites (Manjinggou)	1817 ± 12	SHRIMP	Guo et al. (2005)
08MQG22	Meta-zircons from Al-rich gneiss	1853 ± 15	ICP-MS	Wang et al. (2010a)
08MQG23	Meta-zircons from Al-poor gneiss	1842 ± 10	ICP-MS	Wang et al. (2010a)
08MQG26	Meta-zircons from HP mafic granulite	1839 ± 22	ICP-MS	Wang et al. (2010a)
08MQG28	Meta-zircons from TTG gneiss	1846 ± 21	ICP-MS	Wang et al. (2010a)
08MQG28	Igneous zircons from leucosomes of TTG gneiss	1819 ± 13	ICP-MS	Wang et al. (2010a)
08MQG30	Meta-zircons from TTG gneiss	1853 ± 28	ICP-MS	Wang et al. (2010a)
08MQG29	Zircons from a pegmatite dyke	1806 ± 15	ICP-MS	Wang et al. (2010a)
M21	Meta-zircons from a tonalitic gneiss	1847 ± 17	SHRIMP	Zhao et al. (2008a)
M19	Meta-zircons from a trondhjemitic gneiss	1842 ± 9.8	SHRIMP	Zhao et al. (2008a)
M23	Meta-zircons from a granodioritic gneiss	1847 ± 11	SHRIMP	Zhao et al. (2008a)
M17	Meta-zircons from a HP mafic granulite	1848 ± 19	SHRIMP	Zhao et al. (2008a)
M28	Meta-zircons from a granitic gneiss	1839 ± 46	SHRIMP	Zhao et al. (2008a)
M22	Igneous zircons from anatectic charnockite	1849 ± 9.8	SHRIMP	Zhao et al. (2008a)
M24	Meta-zircons from a S-type granite	1850 ± 17	SHRIMP	Zhao et al. (2008a)
01M20	Outer rims of metamorphic zircon from khondalite	1850 ± 5	SHRIMP	Zhao et al. (2010a)
06M02	Outer rims of metamorphic zircon from khondalite	1857 ± 16	SHRIMP	Zhao et al. (2010a)

(Continued)

Table 4.2 (Continued)

Sample	Description	Age (Ma)	Method	Sources
08YK	Igneous zircons from hyalophane pegmatite	1812 ± 5	Cameca	Qu et al. (2012)
05LF177	Meta-zircons from dioritic gneiss	1785 ± 14	LA-ICP-MS	Liu et al. (2009b)
MJ37	Meta-zircons from tonalitic gneiss	1848 ± 17	SHRIMP	Liu et al. (2012a)
05LF96	Meta-zircons from trondhjemitic gneiss	1864 ± 14	CAMECA	Liu et al. (2012a)
05LF77	Dioritic gneiss	1832 ± 22	CAMECA	Liu et al. (2012a)
05LF51	Meta-zircons from dioritic gneiss	1854 ± 28	CAMECA	Liu et al. (2012a)
Hengshan Complex				
990803	One euhedral zircon from a dioritic gneiss (Dashiyu)	1881 ± 8	SHRIMP	Kröner et al. (2005b)
M068	Meta-zircons from HP granulite (Dashiyu)	1850 ± 3	SHRIMP	Kröner et al. (2006)
HG2	Meta-zircons from HP granulite (Dashiyu)	1867 ± 23	SHRIMP	Kröner et al. (2006)
Ch990839	Coarse-grained pegmatitic melt (Xiaoshiyu)	1851 ± 5	SHRIMP	Kröner et al. (2006)
	Same sample	1856.1 ± 0.6	SGE	Kröner et al. (2006)
HG1	Meta-zircons from a granitic gneiss (Dashiyu)	1872 ± 17	SHRIMP	Kröner et al. (2006)
Ch980871	Meta-zircons from a retro-eclogite (Xiaoshiyu)	1881.3 ± 0.4	SGE	Kröner et al. (2006)
Ch990853	Meta-zircons from a retrograded eclogite (Dashiyu)	1859.7 ± 0.5	SGE	Kröner et al. (2006)
Ch990886	Meta-zircons from a mafic granulite (Dashiyu)	1850.9 ± 0.4	SGE	Kröner et al. (2006)
Ch990848	Meta-zircons from HP granulite (Dashiyu)	1885.6 ± 0.4	SGE	Kröner et al. (2006)
09LYK03	Meta-zircons from tonalitic gneiss	1831 ± 33	LA-ICP-MS	Zhao et al. (2010b)
09LYK13	Metamorphic zircon from biotite monzogranite	1860 ± 19	CAMECA	Zhao et al. (2010b)
09LYK06	Metamorphic zircon from biotite monzogranite	1854 ± 34	LA-ICP-MS	Zhao et al. (2010b)
FP52	Meta-zircons from migmatiteic leucosome	1850 ± 5	LA-ICP-MS	Faure et al. (2007)
FP35	Monazites from pelitic schist	1883 ± 11	EPMA	Faure et al. (2007)

(Continued)

Table 4.2 (Continued)

Sample	Description	Age (Ma)	Method	Sources
H29	Monazites from metapelite	1884 ± 11	EPMA	Trap et al. (2007)
FP40	Muscovite from granitic dyke	1804 ± 13	Ar/Ar	Trap et al. (2012)
FP35	Biotite from gneissic metapelite	1812 ± 13	Ar/Ar	Trap et al. (2012)
FP33	Biotite from the Yixiangzhai orthogneiss	1855 ± 10	Ar/Ar	Trap et al. (2012)
Wutai Complex				
unkown	Garnet amphibolite from the Jingangku Formation	1851 ± 9	Sm–Nd	Wang et al. (2001)
S2010-2-1	Meta-monazites from a kyanite schist (Jingangku)	1822 ± 14	EPMA	Liu et al. (2006)
S2010-2-1	Meta-monazites from a kyanite schist (Jingangku)	1833 ± 8	EPMA	Liu et al. (2006)
SZ10	Meta-monazites from a kyanite schist (Jingangku)	1847 ± 62	EPMA	Liu et al. (2004b)
W109	Monazites from metapelitie	1887 ± 4	EPMA	Trap et al. (2007)
W175	Monazites from metapelite	1886 ± 5	EPMA	Trap et al. (2007)
FP108	Biotite from garnet-bearing gneiss	1802 ± 13	Ar/Ar	Trap et al. (2012)
FP185	Biotite from meta-rhyolite	1799 ± 13	Ar/Ar	Trap et al. (2012)
FP101	Biotite from the Ekou orthogneiss	1810 ± 13	Ar/Ar	Trap et al. (2012)
FP91	Biotite from mica-schist	1873 ± 12	Ar/Ar	Trap et al. (2012)
Fuping Complex				
FG1	Meta-zircons from a tonalitic gneiss (Xicaokou)	1802 ± 43	SHRIMP	Zhao et al. (2002c,d)
FP217	Meta-zircons from a trondhjemitic gneiss	1875 ± 43	SHRIMP	Zhao et al. (2002c)
FP216	Meta-zircons from granodioritic gneiss (Tuanpokou)	1825 ± 12	SHRIMP	Zhao et al. (2002c)
FP249	One meta-zircon rim from a paragneiss (Jiaan)	1821 ± 42	SHRIMP	Zhao et al. (2002c)
FP260	One meta-zircon rim from a paragneiss (Diaoyutai)	1891 ± 6	SHRIMP	Zhao et al. (2002c)
FP188-2	Meta-zircons from a Nanying granitoid gneisses	1826 ± 12	SHRIMP	Zhao et al. (2002c)
FP204	Meta-zircons from a Nanying granitoid gneisses	1850 ± 9.6	SHRIMP	Zhao et al. (2002c)
FP30	Two meta-zircon rims from a gneissic granite	1825 ± 18	SHRIMP	Guan et al. (2002)

(Continued)

Table 4.2 (Continued)

Sample	Description	Age (Ma)	Method	Sources
FP08	Meta-zircons from a granodioritic gneiss (Xicaokou)	1817 ± 26	SHRIMP	Guan et al. (2002)
FP154	Muscovite from gneiss	1830 ± 12	Ar/Ar	Trap et al. (2008)
FP133	zircons from leucosome in migmatite	1847 ± 7	LA-ICP-MS	Trap et al. (2008)
FP133	Monazites from leucosome in migmatite	1837 ± 6	EPMA	Trap et al. (2008)
FP135	Meta-zircons from migmatite leucosome	1842 ± 12	LA-ICP-MS	Trap et al. (2008)
FP014	Meta-zircons from paragneiss	1820 ± 20	LA-ICP-MS	Xia et al. (2006c)
FP014	Meta-zircons from paragneiss	1863 ± 30	LA-ICP-MS	Xia et al. (2006c)
Zanhuang Complex				
99JX-91	Biotite from a mylonitic gneiss	1826.83 ± 0.8	Ar/Ar	Wang et al. (2003)
99JX-91	Biotite from a mylonitic gneiss	1792.61 ± 2.2	Ar/Ar	Wang et al. (2003)
FP400	Monazites from mica-schist	1824 ± 6	EPMA	Trap et al. (2009a)
FP403	Hornblende from migmatitic orthogneiss	1801 ± 12	Ar/Ar	Trap et al. (2008)
HB135	Meta-zircons from kyanite–garnet gneiss	1821 ± 16	SHIRMP	Xiao et al. (2011b)
Lüliang Complex				
IL03-2-1	Monazite from garnet-bearing felsic gneiss	1847 ± 7	EPMA	Liu et al. (2006)
IL03-2-2	Monazite from garnet-bearing felsic gneiss	1880 ± 9	EPMA	Liu et al. (2006)
IL02-4	Monazite from granitic vein	1867 ± 4	EPMA	Liu et al. (2006)
M112	Meta-zircon from tonalitic gneiss	1872 ± 7	SHRIMP	Zhao et al. (2008b)
FP276	Monazites from kyanite-rich gneiss	1881 ± 10	EPMA	Trap et al. (2009b)
FP313	Monazites from sillimanite-rich gneiss	1913 ± 10	EPMA	Trap et al. (2009b)
FP359	Monazites from sillimanite + garnet metapelite	1887 ± 8	EPMA	Trap et al. (2009b)
FP359	Monazites from sillimanite + garnet paragneiss	1872 ± 12	EPMA	Trap et al. (2009b)

(*Continued*)

Table 4.2 (Continued)

Sample	Description	Age (Ma)	Method	Sources
Taihua Complex				
TW0006/1	Meta-zircons from garnet–sillimanite gneiss	1844 ± 66	SHRIMP	Wan et al. (2006b)
TWJ358/1	Meta-zircons from garnet gneissic granitoid	1871 ± 14	SHRIMP	Wan et al. (2006b)
LS0417-1	Meta-zircons from 2.83 Ga banded amphibolite	2651 ± 13 2792 ± 11	SHRIMP	Liu et al. (2009a)
LS0417-3	Meta-zircons from 2.84 Ga gneissic amphibolite	2671 ± 25 2776 ± 20	SHRIMP	Liu et al. (2009a)
LS0417-2	Meta-zircons from 2.83 Ga tonalitic gneiss	2772 ± 22	SHRIMP	Liu et al. (2009a)
LS0417-4	Meta-zircons from 2.83 Ga tonalitic gneiss	2772 ± 17	SHRIMP	Liu et al. (2009a)
Ages of Syn- or Post-Collisional Granites				
Huai'an Complex				
M22	Igneous zircons from anatectic charnockite	1849 ± 9.8	SHRIMP	Zhao et al. (2008a)
08MQG28	Igneous zircons from leucosomes of TTG gneiss	1819 ± 13	ICP-MS	Wang et al. (2010a)
Lüliang Complex				
M101	Huijiazhuang gneissic granite	1832 ± 11	SHRIMP	Zhao et al. (2008b)
M122	Luyashan charnockite	1815 ± 5	SHRIMP	Zhao et al. (2008b)
M128	Caolugou porphyritic granite	1807 ± 10	SHRIMP	Zhao et al. (2008b)
M104	Guandishan massive granite	1798 ± 11	SHRIMP	Zhao et al. (2008b)
M130	Tangershan massive granite	1790 ± 14	SHRIMP	Zhao et al. (2008b)
FP285	Granite	1873 ± 31	LA-ICP-MA	Trap et al. (2009b)
LL401-1	Gneissic monzogranite	1870 ± 29	SHRIMP	Liu et al. (2009c)
LL401-1	Massivegarnet-bearing monzogranite	1840 ± 27	SHRIMP	Liu et al. (2009c)
Ages of Postorogenic (1780–1750 Ma) Mafic Dike Swarms				
Hengshan Complex				
GU12	Unmetamorphosed mafic dyke (near Tuling)	1769.1 ± 2.5	SGD	Halls et al. (2000)
Wutai Complex				
03WT08	Unmetamorphosed diabase dyke (Wutai County)	1754 ± 71	SGD	Peng (2005)

(Continued)

Table 4.2 (Continued)

Sample	Description	Age (Ma)	Method	Sources
Fuping Complex (Taihangshan)				
99JX-16	Unmetamorphosed mafic dyke (Canyansi)	1765.3 ± 1.1	W−Ar/Ar	Wang et al. (2004b)
99JX-65	Unmetamorphosed mafic dyke (Huangbeiping)	1774.7 ± 0.7	W−Ar/Ar	Wang et al. (2004b)
99JX-71	Unmetamorphosed mafic dyke (Hujian'an)	1780.7 ± 0.5	W−Ar/Ar	Wang et al. (2004b)

SHRIMP, sensitive high-resolution ion microprobe; CAMECA, cameca ion microprobe; SGE, single grain evaporation; SGD, single grain dissolution; EPMA, electron probe microanalysis; LA-ICP-MS, laser-ablation induced coupled plasma-mass spectrometry; Sm−Nd, mineral Sm−Nd; Ar/Ar, mineral $^{40}Ar/^{39}Ar$; Meta-zircon, metamorphic zircons.

terranes, represented by the Wutai and Shushui granitoids; (iii) 2530−2520 Ma greenstone assemblages, represented by the Wutai "Group"; (iv) 2520−2475 Ma high-grade TTG gneisses represented by the Huai'an, Hengshan, and Fuping gneisses; (v) Paleoproterozoic granitoid gneisses including 2360, ~2250, and 2110−2176 Ma gneisses from the Hengshan Complex, 2176−2110 Ma Chijianling gneisses from the Lüliang Complex, and ~2050 Ma Nanying gneisses from the Fuping Complex; (vi) Paleoproterozoic sedimentary-volcanic successions represented by the Hutuo, Lüliang, Yejishan, Gantiaohe, Zhongitiao, and Songshan groups; (vii) 2190−1920 Ma mafic dykes that were metamorphosed to amphibolites and medium- to high-pressure mafic granulites; and (viii) 1.80−1.75 Ga postorogenic or anorogenic mafic dykes (Peng, 2005; Peng et al., 2005, 2007).

Geochemical data suggest that Neoarchean and Paleoproterozoic high-grade TTG gneisses and low-grade granitoids and greenstones in the TNCO developed under continental magmatic arc, island arc, or back-arc basin environments (Bai et al., 1992; Sun et al., 1992b,c; Liu et al., 2000b, 2002a,b, 2004a, 2005, 2009c, 2012d; Wang et al., 2004a, 2010b,c; Huang et al., 2010), and minor ultramafic to mafic rocks from the granite-greenstone terranes have been interpreted as remnants of ancient oceanic crust, as represented by the peridotites, gabbros, and pillow basalts from the Wutai and Lüliang complexes (Li et al., 1990; Wang et al., 1996, 1997; Polat et al., 2005), whereas the Paleoproterozoic sedimentary-volcanic succession are considered to have developed in continental rift basins (Zhao et al., 1993; Miao et al., 1999) or foreland basins (Li and Kusky, 2007; Faure et al., 2007; Trap et al., 2007, 2008, 2009a; Liu et al., 2011a,b, 2012b,c).

The most important lithologies in the TNCO are retrograded eclogites and high-pressure mafic and pelitic granulites (Zhai et al., 1992, 1995; Guo et al., 1993, 2001, 2002, 2005; Guo and Shi, 1996; Ma and Wang, 1995; Guo and Zhai, 2001; Zhao et al., 2001b, 2006; Zhang et al., 2006a), which are considered as robust evidence for collision tectonics (Zhao et al., 2001a, 2005). High-pressure mafic granulites in the TNCO were first recognized as enclaves in granitoids from the Hengshan Complex by Wang et al. (1991b) and as boudins or sheets in the Huai'an Complex by Zhai et al. (1992), and later, Zhai et al. (1995) discovered retrograded eclogites from the Sanggan tectonic belt that is situated between the Hengshan and Huai'an complexes. Considering the presence of retrograded eclogites and high-pressure mafic granulites in the Huai'an Complex, Zhai et al. (1992, 1995), at that time, interpreted the Sanggan tectonic belt as a continent–continent collisional belt along which the Huai'an Block in the north and the Hengshan Block in the south collided to form a coherent block. Later, similar high-pressure mafic granulites and retrograded eclogites were also found in the Hengshan, Xuanhua, and Chengde complexes (Figure 4.12; Guo et al., 1993, 2002, 2005; Guo and Shi, 1996; Li et al., 1998; Mao et al., 1999; Zhao et al., 2001b). In the field, the retrograded eclogites and high-pressure mafic granulites outcrop as enclaves, boudins and sheets, ranging from 0.1 to 2 m in width and from 0.1 to 50 m in length, within heterogeneous, migmatitic, and deformed upper amphibolite to granulite-facies TTG gneisses (Figure 4.13A and B). In some low-strain zones, the high-pressure granulites occur as dykes which can be traced up to several hundred meters in length (Figure 4.13C). Magmatic and metamorphic zircons from these high-pressure granulite-facies mafic dykes yielded SHRIMP ages of 2190–1920 and ~1850 Ma, respectively (Kröner et al., 2006). On the outcrop scale, it can be observed that coarse-grained garnet porphyroblasts in the high-pressure granulite are surrounded by symplectic plagioclase and pyroxene (Figure 4.13D). Ma and Wang (1995) discovered high-pressure pelitic granulites from the Xuanhua Complex, which is situated in the northern sector of the TNCO. High-pressure pelitic granulites are generally regarded as the earmark of subduction/collision belts because only subduction/collision processes could bring the sedimentary protoliths of pelitic granulites down to a lower crustal

Figure 4.13 Occurrences of high-pressure basic granulites from the Hengshan Complex in the Trans-North China Orogen (Zhao, 2009). (A) High-pressure basic granulite boudins in TTG gneisses; (B) Discontinuous distribution of high-pressure basic granulite lenses along foliations; (C) Gabbroic dykes of high-pressure basic granulite; (D) 'White-eye-socket' textures in high-pressure basic granulite.

level where they experienced high-pressure granulite-facies metamorphism. Taken together, high-pressure mafic and pelitic granulites and retrograded eclogites constitute a northeast-southwest trending zone that extends from the Hengshan Complex, through the Huai'an and Xuanhua complexes, into the Chengde Complex for a distance of ~500 km in the northern sector of the TNCO (Figure 4.12), whereas the southern segment of the orogen contains ca. 10−14 kbar garnet amphibolites and kyanite−staurolite−anthophyllite mafic schists in the Wutai, Zanhuang, Lüliang, and Zhongtiao complexes (Figure 4.12; Wang et al., 1996, 1997; Zhao et al., 1999c, 2000a,b; Xiao et al., 2011a,b). Therefore, high-pressure rocks crop out along the whole segment of the TNCO.

4.4.2 Metamorphic Evolution of the TNCO

Most medium- and high-grade metamorphic rocks in the TNCO preserve the peak, post-peak decompression and later cooling mineral assemblages (Zhao et al., 2000a and references wherein). For example, in the retrograded eclogites in the Hengshan Complex, the peak (M1) assemblage is omphacite pseudomorph (clinopyroxene + Na-rich plagioclase) + garnet + quartz, whereas the decompression (M2) assemblage is represented by symplectite clinopyroxene + Na-rich plagioclase, which still remains a crystal shape of omphacite (Figure 4.14A) and thus is interpreted as the breakdown of omphacite during the transformation of eclogites to high-pressure mafic granulites. A later cooling mineral assemblage (M3) in the retrograded eclogites is the hornblende + plagioclase symplectite surrounding garnet grains. In the high-pressure mafic granulites from the Hengshan, Huai'an, Xuanhua, and Chengde complexes, the peak (M1) assemblage is garnet + quartz + plagioclase + clinopyroxene; the post-peak decompression (M2) assemblage is the orthopyroxene + Ca-rich plagioclase

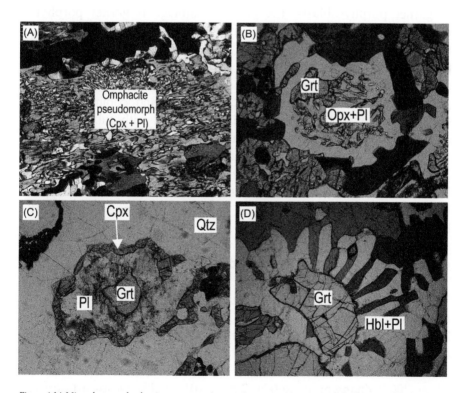

Figure 4.14 Microphotographs showing representative reaction textures in retrograded eclogites and high-pressure mafic granulites from the TNCO (Zhao et al., 2000a, 2001b): (A) omphacite pseudomorph (clinopyroxene + Na-rich plagioclase); (B) orthopyroxene + Ca-rich plagioclase symplectite; (C) clinopyroxene + plagioclase corona; (D) hornblende + plagioclase symplectite. Cpx, clinopyroxene; Grt, garnet; Hbl, hornblende; Opx, orthopyroxene; Pl, plagioclase.

symplectite (Figure 4.14B) or the clinopyroxene + plagioclase corona (Figure 4.14C) surrounding the embayed garnet grains, and the later retrogressive assemblage is the hornblende + plagioclase symplectite around garnet (Figure 4.14D). In the medium-pressure mafic granulites, the peak assemblage is garnet + plagioclase + clinopyroxene + orthopyroxene, whereas the post-peak decompression and later cooling mineral assemblages are the same as those in the high-pressure mafic granulites. In the high-pressure pelitic granulites from the Xuanhua Complex, the peak assemblage is kyanite + K-feldspar + garnet + biotite + plagioclase + quartz, and the post-peak decompression assemblages are represented by the formation of the sillimanite + cordierite symplectite replacing the matrix-type biotite and the cordierite corona surround the embayed garnet (Ma and Wang, 1995). In amphibolites, the prograde (M1) assemblage is plagioclase + quartz + actinolite + chlorite + epidote + biotite + rutile, preserved as mineral inclusions in garnet porphyroblasts, the peak assemblage is garnet porphyroblast + quartz + plagioclase + amphibole + biotite + rutile + ilmenite, and the decompression assemblage is the hornblende/cummingtonite + plagioclase symplectite surrounding the embayed garnet grains. These mineral assemblages and their $P-T$ estimates define clockwise $P-T$ paths involving near-isothermal decompression and cooling following the peak metamorphism (Figure 4.15), consistent with subduction and continent–continent collision environments (Zhao et al., 2000a and references wherein). This formed an initial justification for Zhao et al. (1998, 2001a) to have proposed the threefold division of the NCC into the Eastern and Western blocks and intervened TNCO that was a typical Himalayan-type continent–continent collisional belt along which the two blocks collided to form the NCC.

4.4.3 Structures of the TNCO and Subduction Polarity

There is a hot debate on the polarity of subduction operative in the TNCO, with many researchers arguing for westward-directed subduction models (Kusky and Li, 2003; Li and Kusky, 2007; Faure et al., 2007; Santosh, 2010; Kusky, 2011a,b; Deng et al., 2013), whereas others favor eastward-directed subduction models (Zhao et al., 2001a, 2005; Wilde et al., 2002; Kröner et al., 2005a,b, 2006; Wilde and Zhao, 2005; Zhang et al., 2007a, 2009a, 2012c). The westward-directed subduction models are based mainly on two lines of evidence: (i) the top-to-the-SE thrusting and shearing structures in the TNCO (Trap et al., 2007, 2008, 2009a,b, 2011, 2012) and (ii) some seismic images of

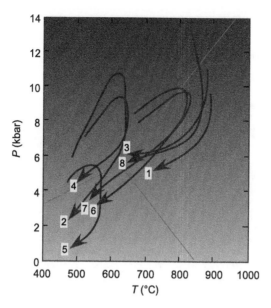

Figure 4.15 P–T paths of metamorphic rocks in the TNCO (Zhao et al., 2000a and references wherein). 1, Mafic granulites from the Hengshan complex (Zhao et al., 2001b); 2, amphibolite from the Wutai Complex (Zhao et al., 1999c); 3, mafic granulites from the Fuping Complex (Zhao et al., 2000b); 4, garnetiferous amphibolites from the Lüliang Complex (Zhao et al., 2000a); 5, garnetiferous pelitic rocks from the Zhongtiao Complex (Mei, 1994); 6, pelitic granulites from the Huaian Complex (Liu, 1995); 7, mafic granulites from the Huai'an Complex (Liu, 1995); 8, high-pressure granulites from the Hengshan Complex (Zhao et al., 2001b).

the NCC implying relict west-dipping slabs (Chen et al., 2008, 2009; Zheng et al., 2009b).

However, recent structural data show that not all thrusting and shearing structures in the TNCO are the top-to-the-southeast. Zhang et al. (2007a, 2009a, 2012c) and Li et al. (2010a) carried out detailed structural investigations on the Fuping, Wutai, and Hengshan complexes, which constitute a representative transect across the central segment of the TNCO (Figure 4.12) This region consists of two high-grade gneiss complexes (Fuping and Hengshan complexes) in the southeast and northwest, respectively, separated by the low-grade Wutai granite-greenstone terrane (Figure 4.16). The Fuping Complex in the southeast is dominated by a top-to-the-SE(E) structures (Zhang et al., 2009a), whereas the Hengshan Complex is characterized by a top-to-the-NW sense of thrusting and shearing structures (Zhang et al., 2007a; Li et al., 2010a). The intervening Wutai Complex has similar structural features to those of the Fuping Complex in its southeastern part and to those of the Hengshan Complex in its northwestern part

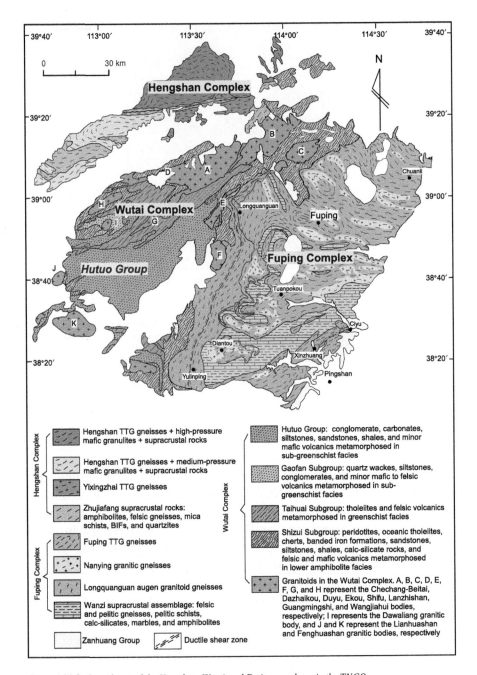

Figure 4.16 Geological map of the Hengshan, Wutai, and Fuping complexes in the TNCO.

and (Zhang et al., 2012c; Li et al., 2010a). Therefore, the overall structural pattern of the Fuping–Wutai–Hengshan region is fan-shaped, with opposite thrusting geometries where the thrusting and shearing is the top-to-the-SE(E) in the southeastern and top-to-the-NW in the northwestern (Zhang et al., 2012c; Li et al., 2010a). In fact, many classic Phanerozoic continent–continent collisional belts are also characterized by fan-shaped or flower-shaped structural patterns, including the Himalayas (Ziegler et al., 1995), the Alps (Escher and Beaumont, 1997), and the Appalachians (Castonguay and Tremblay, 2003). Theoretically, neither the eastward-directed nor westward-directed subduction is a unique model for such a "fan-shaped structural pattern" in the TNCO. Thus, the subduction polarity of the TNCO cannot be determined simply based on the senses of thrusting and shearing structures, but requires an integrated structural, metamorphic, geochronological, geochemical, and geophysical study.

Recently, a number of researchers have carried out seismic imaging work on the NCC (Chen et al., 2008, 2009; Tian et al., 2009; Zheng et al., 2009b, 2010, 2012; Zhu and Zheng, 2009; Zhao et al., 2011b; Chang et al., 2012; Cheng et al., 2012; Jiang et al., 2012), though few researchers applied their data to resolve Precambrian issues. Based on seismic images along a profile across the TNCO and the Western Block, Zheng et al. (2009b) recognized two low-velocity layers marked as L1 and L2 (Figure 4.17A and B), and interpreted them as westward-directed subducted slabs (Figure 4.17C), consistent respectively with the Trans-North China and Taihangshan sutures as proposed by Faure et al. (2007) and Trap et al. (2009a,b, 2011). However, Zhao et al. (2010c) questioned this interpretation because both L1 and L2 are restricted a crustal depth, not extending into the mantle or offsetting the Moho (e.g., Zheng et al., 2012), and thus whether they represent subducted slabs or overthrust crustal slices formed by exhumation or uplift following the collision between the Western and Eastern blocks along the TNCO still remains unknown. Moreover, layer L2 is nearly horizontal (Figure 4.17B), and thus the direction of subduction inferred by L2 is equivocal. Furthermore, as shown in Figure 4.17D, layer L1 is not located within the TNCO, but situated in the area of the Western Block close to the Khondalite Belt. Therefore, even if L1 represents a subducted slab, it may reflect the subduction structure of the Khondalite Belt, and not that of the TNCO (Zhao et al., 2010c).

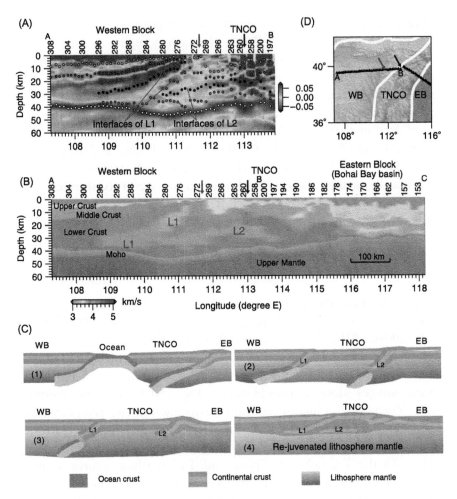

Figure 4.17 Seismic images along a profile across the TNCO and the Western Block (Zheng et al., 2009b). (A) Common conversion point (CCP) receiver function image of crust and uppermost mantle along Western block–TNCO profile based on the inverted velocity model. Dots in CCP image mark velocity discontinuities in the best-fitting models, including upper-middle (orange) and middle-lower (red) crustal interfaces, Moho (white), the interfaces with negative velocity gradient above L1 and L2 layers (blue), and the bottom interfaces of L1 and L2 (green). (B) Shear-wave velocity structure of crust and uppermost mantle compiled from inverted velocity model along east-west profile. L1 is a westward-dipping low-velocity zone beneath stations 274–296 that separates TNCO and Western block, and L2 is a horizontal low-velocity zone in the lower crust beneath TNCO and the Western block. (C) Schematic diagram of ancient subduction model. (D) Topographic map of the study region and locations of stations. Selected station numbers are labeled on top of plots in (A) and (B). In (A), (B), and (D), red arrow marks boundary location between Western block and TNCO identified by Zheng et al. (2009b), and blue arrow marks boundary previously envisaged by Zhao et al. (2001a). EB, Eastern Block; WB, Western Block.

On the other hand, Santosh (2010) reinterpreted the seismic images of Chen et al. (2009) and proposed that the present west-dipping shape of the lithosphere–asthenosphere boundary (LAB) across the TNCO formed by west-directed subduction in the Paleoproterozoic. In their original publication, however, Chen et al. (2009) interpreted the current structures of the LAB in the NCC as a result of the Mesozoic to Cenozoic lithospheric thinning/destruction. Such an interpretation is consistent with recent seismic data which show that the structures of lithospheric crust and mantle in both the Eastern Block and the TNCO were significantly reworked/replaced during Mesozoic and Cenozoic times (Zhao et al., 2011b; Chang et al., 2012; Cheng et al., 2012; Jiang et al., 2012). Thus, it is unlikely that the polarity of Paleoproterozoic subduction operative in the TNCO can be determined from the orientation of the LAB in the NCC.

Zhao et al. (2001a, 2005, 2012) favor an eastward subduction model because the TNCO contains 2.65–2.85 Ga old continental crustal components that are similar in composition and age to those rocks in the Eastern Block, which led them to propose that the TNCO developed on an Andean-type continental margin arc on the western margin of the Eastern Block. In the Xuanhua Complex, Jiang et al. (2010) obtained a LA-ICP-MS U–Pb zircon age of 2715 ± 7 Ma for a pyroxenite xenolith in the Hannuoba basalt. In the Huai'an Complex (Figure 4.12), Liu et al. (1997b) found a large number of xenocrystic zircons with SHRIMP $^{207}Pb/^{206}Pb$ apparent ages from 2761 to 2613 Ma from the ~2500 Ma tonalitic gneisses. In the Hengshan Complex (Figure 4.12), a grey trondhjemitic gneiss sample and a grey biotite gneiss sample were dated by Kröner et al. (2005b) using the SHRIMP at 2712 ± 2 and 2700 ± 6 Ma, respectively. In the Wutai Complex (Figure 4.12), four xenocrystic zircons from the 2553 ± 8 Ma Lanzhishan granites yielded a weighted SHRIMP $^{207}Pb/^{206}Pb$ age of 2702 ± 10 Ma (Wilde et al., 1997). Wilde (2002) also found xenocrystic zircons with age populations of 2763 ± 10 and 2660 ± 7 Ma from the 2534 ± 10 Ma Longquanguan augen gneiss. In addition, two xenocrystic zircons from the 2529 ± 10 Ma meta-andesite of the Zhuangwang Formation yielded a weighted SHRIMP $^{207}Pb/^{206}Pb$ age of 2702 ± 10 Ma (Wilde et al., 2004b). In the Fuping Complex (Figure 4.12), medium-grained hornblende gneiss enclosed in the 2520 Ma Fuping granitoid gneisses gave a SHRIMP U–Pb zircon age of 2708 ± 8 Ma (Guan et al., 2002), and the Paleoproterozoic Wanzi supracrustal rocks contains detrital zircons with ages between 2827 ± 8 and 2686 ± 8 Ma (Zhao et al., 2002c). In the Zhongtiao Complex (Figure 4.12), a xenocrystic zircon from the meta-rhyolitic tuff of the Paleoproterozoic Jiangxian Group gave a SHRIMP $^{207}Pb/^{206}Pb$ apparent age of 2776 ± 16 Ma (Sun et al., 1992c). In the low-grade Dengfeng granite-greenstone terrane (Figure 4.12), the ages of xenocrystic zircons from the 2512 ± 12 Ma meta-dacite of the Dengfeng Group range from 2576 ± 9 to 2945 ± 44 Ma (Kröner et al., 1988). In the high-grade Taihua Complex of the TNCO (Figure 4.12), for example, magmatic zircons yielded weighted SHRIMP $^{207}Pb/^{206}Pb$ ages of 2829 ± 18 and 2832 ± 11 Ma for two tonalitic gneiss samples and 2838 ± 35 and 2845 ± 23 Ma for two amphibolites samples, whereas metamorphic zircons from these samples yielded two age populations: $2772-2792$ and $2638-2671$ Ma (Liu et al., 2009a). Thus, $2650-2850$ Ma old continental components have been reported from nearly all metamorphic complexes in the TNCO. So far, $2650-2850$ Ma rocks have not been reported from the Western (Ordos) Block, but are widely exposed in the Eastern Block (e.g., Taishan Complex; Wan et al., 2011a). In particular, Wan et al. (2011a) reported that, like the ~ 2.8 Ga TTG gneisses of the Taihua Complex in the TNCO, the ~ 2.8 Ga TTG gneisses of the Taishan Complex in the Eastern Block also experienced a ~ 2.65 Ga metamorphic event. In addition, Hf isotopic data also show that the rocks of the TNCO have affinities to those of the Eastern Block, but are in contrast with those of the Western Block (Xia et al., 2009). Taken together, all these suggest that the subduction operative in the TNCO was most likely eastward, forming an Andean-type continental margin arc on the western margin of the Eastern Block.

4.4.4 Timing of Collision Between the Western and Eastern Blocks

The timing of collision between the Eastern and Western blocks along the TNCO to form a coherent basement of the NCC was once hotly debated, with some researchers arguing that it occurred at ~ 2.5 Ga (Kusky and Li, 2003; Li and Kusky, 2007; Kusky et al., 2007; Kusky, 2011a,b), whereas others believed that it occurred at ~ 1.85 Ga (Zhao et al., 2001a, 2005; Wilde et al., 2002; Guo et al., 2005; Kröner et al., 2005a,b, 2006; Liu et al., 2006). However, now it is no longer a controversial issue since so much microanalysis dating work has been done on metamorphic zircons from various high-grade metamorphic rocks in the TNCO in the last decade and nearly all metamorphic zircons give consistent ages around 1.85 Ga, without any one around 2.5 Ga (see Table 4.2 and references wherein). Metamorphic mineral Ar/Ar and Sm−Nd ages and EPMA Th−U−Pb monazite ages obtained for the rocks of the TNCO are also around 1.9−1.8 Ga (Wang et al., 2003, 2004b; Guo et al., 2005; Liu et al., 2006), consistent with the ages of metamorphic zircons. Moreover, the metamorphic ages of ~1.85 Ga were not only obtained from the Huai'an, Hengshan, Wutai, Fuping, Zanhuang, and Lüliang complexes in the middle segment of the orogen, but were also produced from the Chengde Complex in the northernmost part (Mao et al., 1999) and the Taihua Complex in the southernmost part of the orogen (Wan et al., 2006b). Thus, it is certain that the 1.85 Ga collisional event occurred along the whole TNCO. It deserves mentioning here that Liu et al. (2009a) obtained the metamorphic ages of ~2.65 Ga for the ~2.8 Ga TTG gneisses from the Taihua Complex, which is invalid to be used as evidence against the ~1.85 Ga collision model because the metamorphic ages of ~1.85 Ga have been obtained for Al-rich gneisses from the same complex (Wan et al., 2006b), and as discussed above, these 2.8−2.7 Ga rocks in the TNCO are interpreted as the remnants of old continental basement from the Eastern Block, as their rock-forming and metamorphic ages are similar to those of the 2.8−2.7 Ga granite-greenstone belt in Western Shandong (Wan et al., 2011a).

Kusky and Li (2003) once proposed that the NCC formed by amalgamation of the Eastern and Western blocks at ~2.5 Ga, but more recently Kusky (2011a) modified his model by suggesting that the amalgamation of the two blocks was not completed until 2.3 Ga (also see Deng et al., 2013). In their models, they interpreted the ~1.85 Ga metamorphic event in the TNCO as reworking/overprinting by collision of the northern margin of the NCC with another continental block in the Columbia (Nuna) supercontinent at 1.92−1.85 Ga (Kusky and Li, 2003; Kusky et al., 2007; Kusky, 2011a). However, these models are supported by data listed in Table 1.1, which show that all metamorphic zircon ages obtained from the TNCO are consistently around 1.85 Ga, neither around 2.5 Ga nor around 2.3 Ga. If the collisional events leading to the amalgamation of the Eastern and Western blocks did occur at ~2.5 or 2.3 Ga, it would be unlikely that all age information of these events have been obliterated by younger overprinting events. In particular, the overprinting model has difficulties in explaining the ~1.85 Ga metamorphic event well to the south of the region of reworking proposed by Kusky et al. (2007). For example, Wan et al. (2006b) obtained metamorphic ages of ~1.85 Ga for high-grade pelitic gneisses form the Taihua Complex, which is located in the southernmost part of the TNCO (Figure 4.12). Moreover, the over-printing model cannot well explain why Paleoproterozoic retrograded eclogites and high-pressure granulites are exposed in the Hengshan Complex situated in the central part of the TNCO (Figure 4.12), but not in the so-called "Inner Mongolia-North Hebei Orogen," which is interpreted as having formed by collision of the northern margin of the NCC with another continental block in the Columbia (Nuna) super-continent at 1.92–1.85 Ga.

An end-Neoarchean (~2.5 Ga) foreland basin, called the "Qinglong foreland basin," was proposed to support the ~2.5 Ga col-lision model (Kusky and Li, 2003; Li and Kusky, 2007). The basin is considered to extend N to NE up to 1600 km long along the eastern side of the central orogenic belt (Kusky and Li, 2003; Li and Kusky, 2007). According to Li and Kusky (2007), this foreland basin can be divisible into the southern, middle, and northern segments, of which the southern segment is represented by the Songshan Group in the Songshan area, the middle segment includes the Gaofan and Doucun groups of the Hutuo Supergroup in the Wutaishan area, and the Wanzi Group in the Taihangshan area, and the northern segment is represented by the Qinglong Group in the East Hebei area (Li and Kusky, 2007). Most recently, however, Liu et al. (2011c,d, 2012b,c, 2012e, 2013c) showed that most of these groups contain 2.0–1.8 Ga detrital zircons, indicating that they were not deposited in at the end-Neoarchean (~2.5 Ga), but formed in a Paleoproterozoic basin.

Taken together, all available data convincingly support that the collision between the Eastern and Western blocks to form the TNCO occurred at ~1.85 Ga, leading to the final amalgamation of the NCC.

4.5 SUMMARY

Figure 4.18 is a cartoon diagram showing the Paleoproterozoic amal-gamation of the NCC, based on available data and constraints out-lined above for the period from pre-2.2 to 1.85 Ga. Before 2.2 Ga, the NCC did not have a uniform basement but consisted of four

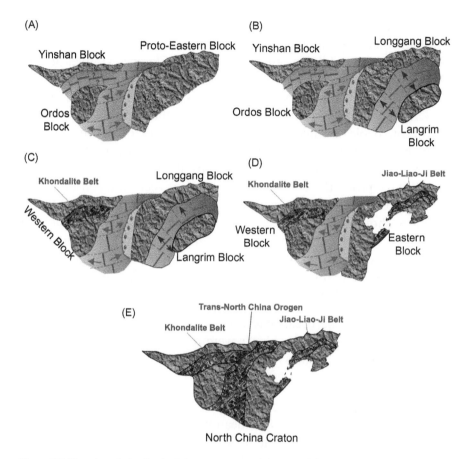

Figure 4.18 Tectonic evolution for the Paleoproterozoic amalgamation of the NCC proposed by Zhao et al. (2012). (A) >2.2 Ga, (B) 2.2–1.95 Ga, (C) ~1.95 Ga, (D) 1.93–1.90 Ga, and (E) ~1.85 Ga.

microcontinental blocks, named the Yinshan and Ordos blocks in the west and the Eastern Block in the east (Figure 4.18A).

In the period 2.2–1.95 Ga, the Archean Eastern Block underwent a rifting event, leading to the development of a rift basin in which granitoid and mafic intrusions were emplaced at 2.2–2.0 Ga and sedimentary-volcanic rocks formed at 2.0–1.95 Ga. The further extension of this rift basin led to the opening of an incipient ocean that broke up the Eastern Block into the Longgang Block in the northwest and the Langrim Block in southeast (Figure 4.18B).

The Yinshan Block is a typical Neoarchean block that was separated from the Ordos Block by an ancient ocean whose lithosphere started to subduct beneath the southern margin of the Yinshan Block at about 2.45 Ga, forming 2435 ± 12 Ma adakites, 2426 ± 41 Ma sanukitoids, and 2416 ± 8 Ma Closepet granites in the Daqingshan–Wulashan area (Zhong, 2010). Although the nature of the Ordos Block is unclear as it is completely covered by Mesozoic–Cenozoic strata, it is inferred that the block was a Paleoproterozoic juvenile arc terrane as the ages of detrital zircons from the khondalite series rocks surrounding the block range between 2.3 and 2.0 Ga, without any Archean detrital zircons in the rocks. The sedimentary protoliths of the khondalite series rocks are considered to have deposited on passive continental margins of the Ordos Block within a short period between 2.0 and 1.95 Ga. At ~ 1.95 Ga, the ocean between the Yinshan and Ordos blocks was closed, leading to collision between the continental arc margin of the Yinshan Block with the passive margin of the Ordos Block, forming the Khondalite Belt and the united Western Block (Figure 4.18C).

At ~ 1.90 Ga, the ocean basin was closed due to subduction, and the Longgang and Langrim blocks were reassembled to form the Jiao-Liao-Ji Belt (Figure 4.18D).

In the period 2.56–1.85 Ga, the western margin of the Eastern Block (Longgang Block in the period 2.2–1.9 Ga) underwent long-lived east-ward subduction, forming an Andean-type continental margin, which was separated from the Western Block (Yinshan and Ordos blocks before ~ 1.95 Ga) by an old ocean. In the period 2550–2520 Ma, the subduction caused partial melting of the medium-lower crust, producing copious granitoid magmas that were intruded into the upper crustal levels to form granitoid plutons in the granite-greenstone terranes. In the period 2530–2520 Ma, the further subduction of the oceanic lithosphere caused partial melting of the mantle wedge, leading to underplating of mafic magma in the lower crust and widespread mafic and minor felsic volcanism in the continental margin arc, forming part of the greenstone assemblages. At 2520–2475 Ma, the subduction caused further partial melting of the lower crust to form large amounts of TTG magmatism. In the period 2.35–2.05 Ga, episodic granitoid magmatism occurred, result-ing in the emplacement of 2360, ~ 2250, 2110–21760, and ~ 2050 Ma granites, probably due to development of some back-arc or intra-arc basins, in which contemporary volcano-sedimentary rocks formed. At 2.05–1.92 Ga, several extensional events occurred, possibly due to subduction of an oceanic ridge, leading to emplacement of mafic dykes that were subsequently metamorphosed to amphibolites and medium- to high-pressure mafic granulites. At 1880–1820 Ma, the ocean was completely consumed by subduction, and the Eastern and Western blocks collided to form the TNCO, leading to formation of the coherent basement of the NCC (Figure 4.18E).

In summary, the NCC was formed by amalgamation of a number of microcontinental blocks in the period 1.95–1.85 Ga, which was coincident with the global 2.1–1.8 Ga collisional events that led to the assembly of the proposed Paleo-Mesoproterozoic Columbia (Nuna) supercontinent.

CHAPTER 5

Mesoproterozoic Accretion and Meso-Neoproterozoic Extension and Rifting of the North China Craton

5.1 INTRODUCTION

There is now a broad agreement that the assembly of the Paleo-Mesoproterozoic Columbia (Nuna) supercontinent was completed by global-scale 2.1−1.8 Ga collisional events (see an overview by Zhao et al., 2002a). Following its final assembly at ∼1.8 Ga, the Columbia supercontinent underwent a long-lived, subduction-related outgrowth along some of its continental margins, forming a number of superlarge accretionary zones, including the huge 1.8−1.3 Ga magmatic accretionary belt bordering the present southern margin of North America, Greenland, and Baltica, the 1.80−1.45 Ga Rio Negro−Juruena Belt and 1,45−1.30 Ga Rondonian−San Ignacio Belt along the western margin of South America, and the 1.8−1.5 Ga Arunta, Musgrave, Mt. Isa, Georgetown, Coen, and Broken Hill inliers along the southern and eastern margins of the North Australian Craton (Figure 5.1; Zhao et al., 2004a). The extension and fragmentation of this supercontinent began about 1.6 Ga ago, in association with the development of Mesoproterozoic continental rifting along the western margin of Laurentia (Wernecke, Muskwa, Belt, Purcell, Uinta, Unkar, and Apache supergroups or groups), southern margin of Baltica (Telemark Supergroup), southeastern margin of Siberia (Riphean aulacogens), and northwestern margin of South Africa (Kalahari Copper Belt). The extension was also associated with widespread anorogenic activity including emplacement of anorthosite−mangerite−charnockite−granite (AMCG) suites and potassic rapakivi granites, especially in North America, Greenland, Baltica, and South America (Rogers and Santosh, 2002; Zhao et al., 2002a, 2004a). The final breakup of the supercontinent is marked by the emplacement of 1.3−1.2 Ga mafic dyke swarms (the McKenzie, Sudbury, Seal Lake, Harp, and Mealy) and coeval eruption of flood basalts (e.g., Coppermine River basalts)

Precambrian Evolution of the North China Craton. DOI: http://dx.doi.org/10.1016/B978-0-12-407227-5.00005-5

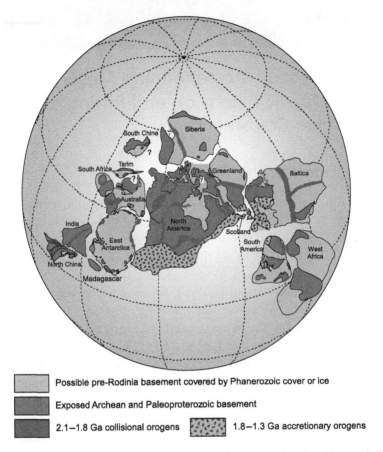

Figure 5.1 The Paleo-Mesoproterozoic Columbia (Nuna) supercontinent, configured by Zhao et al. (2002a, 2004a), showing the distribution of global-scale 2.1–1.8 Ga collisional orogens and 1.8–1.3 Ga accretionary orogens.

in North America and their equivalents in other cratonic blocks (Zhao et al., 2004a).

Also, as discussed in Chapter 4, like most other cratons, the North China Craton (NCC) records the history of the assembly of the Columbia supercontinent. However, little is known about whether the NCC, like North America, Greenland, Baltica, Amazonia, and other cratons, preserves records for the Paleo-Mesoproterozoic outgrowth, extension/fragmentation, and final breakup of the Columbia supercontinent. This chapter summarizes available data and demonstrates that the NCC underwent Mesoproterozoic accretion at its southern margin, Meso-Neoproterozoic extension in its interior, and rifting along its northern margin.

It deserves mentioning here that in the geological timescale of China, the time boundary between Paleoproterozoic and Mesoproterozoic is used to be placed at 1.8 Ga, not at 1.6 Ga.

5.2 MESOPROTEROZOIC ACCRETION ON THE SOUTHERN MARGIN OF THE NCC

Along the southern margin of the NCC is a large Mesoproterozoic volcanic belt, named the Xiong'er volcanic belt (Figure 5.2), which is most likely to have resulted from the outgrowths of the Columbia (Nuna) supercontinent. The belt extends from the Zhongtiaoshan area in the north to the Xiaoshan, Xiong'ershan, and Waifangshan areas in the south (Figure 5.3), covering an area of more than 60,000 km^2. In the south, the belt is separated from the North Qinling orogen by the Luonan–Luanchun Fault, and in the north and west, the belt is bordered on the southern margin of the NCC by the Luoyang–Baofeng and Jiangxian–Lintong faults, respectively (Figure 5.3). The North

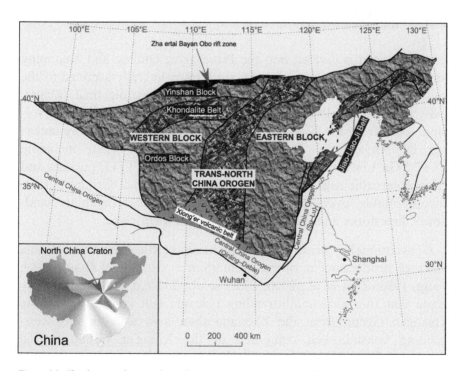

Figure 5.2 Sketch map showing the Paleo-Mesoproterozoic Xiong'er volcanic belt and Mesoproterozoic Zhaertai–Bayan Obo rift zone on the southern and northern margins of the NCC, respectively (Zhao et al., 2003, 2011a).

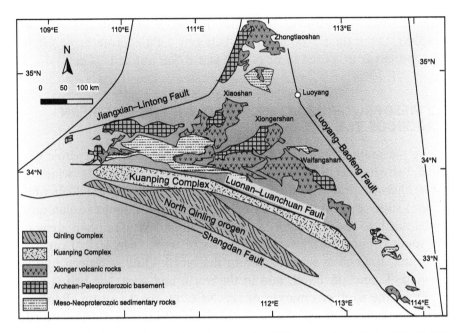

Figure 5.3 Generalized geological map of the southern part of the NCC, showing the distribution of the Xiong'er volcanic belt. Modified from He et al. (2008).

Qinling orogen is divided into the Proterozoic Qinling and Kuanping complexes and is considered as an arc terrane laterally accreted to the southern margin of the NCC during Mesoproterozoic time (Zhang, 1989; Zhang et al., 1995, 1996a,b; Ouyang and Zhang, 1996). The volcanic rocks from the Xiong'er volcanic belt are unconformably underlain by Archean to Paleoproterozoic basement rocks, including 2.8−2.5 Ga tonalitic−trondhjemitic−granodioritic (TTG) gneisses and Paleoproterozoic supracrustal rocks called the Taihua Group, but are unconformably overlain by Meso-Neoproterozoic terrigenous sandstones, limestones and calc-silicate rocks, and themselves.

Conventionally, most researchers interpret the Xiong'er volcanic rocks as products of a rifting event (Sun et al., 1981; Zhang 1989; Yang, 1990; Zhai et al., 2000; Zhao et al., 2002c,d, 2004b), and they argue that the north-south-trending volcanic belt constituted by the Xiyanghe Group from the Zhongtiaoshan area and the east-west-trending volcanic belt composed of the Xiong'er Group in the Xiaoshan, Xiong'ershan, and Waifangshan areas constitute a triple junction rift system (Figure 5.3; Sun et al., 1981; Zhang 1989; Yang, 1990; Zhai et al., 2000). Recently, triple junction rift system is considered to have resulted from a mantle plume event that led to the fragmentation and breakup of the NCC from an unknown cratonic block during the breakup of the Columbia supercontinent (Peng et al., 2005, 2007, 2008; Hou et al., 2008b). The main argument for this model is that the Xiong'er volcanic rocks and the northwest-southeast-trending mafic dyke swarms along the Trans-North China Orogen (TNCO) were nearly coeval. Peng et al. (2008) even speculated that the Xiong'er volcanic rocks and the mafic dyke swarms constitute a Large Igneous Province (LIP) formed by a mantle plume that led to the breakup of the Columbia supercontinent. In this model, Peng et al. (2008) regarded the mafic dykes as the feeders through which the magmas were erupted to the surface to form the Xiong'er volcanic rocks. However, the rift or mantle plume models cannot reasonably explain the petrological, geochemical, and geochronological features of the Xiong'er volcanic rocks, which are consistent with a continental margin arc model (e.g., Chen and Fu, 1992; Chen et al., 1992; He et al., 2008, 2009, 2010a,b).

Volcanic rocks and associated minor sedimentary rocks from the Xiong'er volcanic belt were traditionally named the "Xiong'er Group" in the Xiaoshan, Xiong'ershan, and Waifangshan areas, and the "Xiyanghe Group" in the Zhongtiaoshan area (Figure 5.3; Sun et al., 1981; Zhao et al., 2002c,d, 2004b). These groups are composed of lavas and pyroclastic rocks interlayered with minor sedimentary rocks (<5%), which are subdivided, from bottom to top, into a sedimentary formation, named the Dagushi Formation, and three volcanic formations, named the Xushan, Jidanping, and Majiahe formations (Chen and Fu, 1992; Chen et al., 1992; He et al., 2008, 2009; Zhao et al., 2002c,d, 2004b). Of three volcanic formations, the Xushan Formation comprises basaltic andesites and andesites, with minor dacites and rhyodacites; the Jidanping Formation is composed of dacites, rhyodacites, and rhyolites with minor basaltic andesites and andesites, whereas the Majiahe Formation consists of basaltic andesites, andesites, and voluminous sedimentary and pyroclastic rocks. As a whole, the volcanic rocks in the Xiong'er volcanic belt are composed of basaltic andesites, andesites, dacites, and rhyolites, which are different from bimodal volcanic assemblages within intracontinental rift zones but are remarkably similar to volcanic rock associations on modern continental margin arcs or island arcs. It is impossible that a mantle plume-driven rifting event did not form any basalt, especially in the case that the mafic dyke swarms are regarded as the feeders. Therefore, the rock assemblages of the Xiong'er volcanic belt are not consistent with an interpretation that the Xiong'er volcanic belt represented an intracontinental rift zone formed by a mantle plume-driven rifting event (e.g., Sun et al., 1981; Zhai et al., 2000; Zhao et al., 2002c,d, 2004b; Peng et al., 2005, 2007, 2008; Hou et al., 2008b), but support the model that it formed under an Andean-type continental margin arc or island arc setting (e.g., Chen and Fu, 1992; He et al., 2008, 2009, 2010a,b; Zhao et al., 2009c).

Geochemical data for the Xiong'er volcanic rocks strongly support the continental margin arc model (He et al., 2008, 2010a,b). The Xiong'er volcanic rocks are enriched in LILE (large-ion lithophile elements) and LREE (light rare earth elements), with obviously negative Ta−Nb−Ti anomalies (Figure 5.4), which are common geochemical features of volcanic rock generated by hydrous melting of a mantle wedge in a subduction zone (Pearce and Peate, 1995). The Xiong'er basaltic andesites and andesites are characterized by high Sr/Nd ratios at low Th/Yb ratios, implying that the fluids were originated from the subducting slab (He et al., 2008, 2010a), though the Xiong'er basaltic andesites and andesites have higher Fe−Ti and HFSE (high field strength elements) concentrations than the typical Phanerozoic, intraoceanic arc-related basalts and andesites (Sajona et al., 1996). As shown in Figure 5.5, some of the basaltic andesites and andesites have high TiO_2, P_2O_5, and Nb, plotting in the Nb-enriched basalt field, which requires that the mantle peridotite may have been fluxed by siliceous (adakitic) melts (Sajona et al., 1996). In summary, the major elements, trace elements, and isotopes of the Xiong'er volcanic rocks suggest that they were most likely produced in a subduction zone.

The timing of volcanism to form the Xiong'er volcanic belt has been well constrained recently. Using the sensitive high-resolution ion microprobe (SHRIMP) U−Pb zircon technique, Zhao et al. (2004b) obtained magmatic zircon ages of ∼1.78 Ga for a number of samples collected from volcanic rocks in the Waifangshan area, suggesting volcanism at about 1.78 Ga. This has been confirmed by the SHRIMP and LA-ICP-MS U−Pb zircon ages for volcanic rocks in the Xiyanghe, Xiaoshan, and Xiong'ershan areas (He et al., 2009). For example, four samples collected from different sequences of the Xushan Formation yielded weighted mean $^{207}Pb/^{206}Pb$ ages of

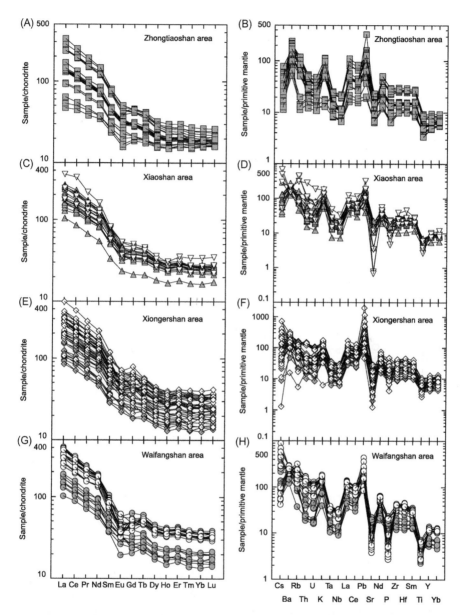

Figure 5.4 Chondrite and primitive mantle normalized plots of the Xiong'er volcanic rocks in the Zhongtiao Mountain (A and B), Xiaoshan (C and D), Xiong'ershan (E and F), and Waifangshan (G and H). Chondrite and primitive mantle values are from Sun and McDonough (1989). Symbols: gray squares, triangles, diamonds, and circles are basaltic andesites and andesites, whereas white symbols are dacites and rhyolites. All data are from He et al. (2008, 2010a,b).

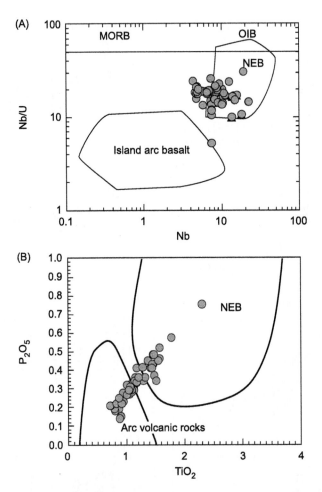

Figure 5.5 (A) Plot of Nb/U versus Nb for the Xiong'er basaltic andesites and andesites. The fields of island arc basalts and Nb-enriched (or High-Nb) arc basalts (NEB) are from Kepezhinskas et al. (1996). (B) Plot of P_2O_5 versus TiO_2 for the Xiong'er basaltic andesites and andesites. The fields of arc volcanic rocks and NEB are from Defant et al. (1992). All data are from He et al. (2008, 2010a,b).

1778 ± 8, 1783 ± 13, 1767 ± 47, and 1783 ± 20 Ma (He et al., 2009). These samples also contain xenocrystic/inherited zircons with ages ranging from 2.55 to 1.91 Ga, similar to the ages of basement rocks on the southern margin of the NCC, suggesting that the Xiong'er volcanic belt represented a continental margin, not an intraoceanic arc (island arc). In addition, He et al. (2009) dated three volcanic samples collected from the Jidanping Formation, of which two rhyolite samples yielded weighted mean [207]Pb/[206]Pb ages of 1778 ± 5.5 and 1751 ± 14 Ma, similar to the ages of the Xushan Formation, but one dacite sample gave a weighted mean $^{207}Pb/^{206}Pb$ age of 1450 ± 31 Ma, which is the youngest age obtained from the volcanic rocks from the Xiong'er volcanic belt. One andesite sample collected from the Majiahe Formation yielded a weighted mean $^{207}Pb/^{206}Pb$ age of 1778 ± 6.1 Ma, interpreted as the age of the Majiahe andesite, coeval with the formation of most volcanic rocks from the Xushan and Jidanping formations. The sample also contains an older group of zircons with a weighted mean $^{207}Pb/^{206}Pb$ age of at 1850 ± 5.9 Ma, interpreted as the age of the xenocrystic/inherited zircons. These SHRIMP and LA-ICP-MS U—Pb zircon data clearly demonstrate that most of the Xiong'er volcanic rocks formed at 1.78—1.75 Ga, with minor felsic volcanic rocks at ~1.45 Ga, indicating that the Xiong'er volcanic rocks on the southern margin of the NCC were not the products of a single magmatic event. This conclusion is also supported by Ren et al. (2000) who obtained single-grain zircon ages of ~1.65 Ga from the volcanic rocks of the Jidanping Formation in the Waifangshan area. In summary, the Xiong'er volcanic rocks erupted intermittently over a protracted interval from 1.78, through 1.76—1.75 and 1.65 Ga, to 1.45 Ga, though the major phase of the volcanism occurred at 1.78—1.75 Ga. Such multiple and intermittent volcanism is inconsistent with a mantle plume-driven rifting event, but is consistent with magmatic events in continental margin arcs.

Taken together, the Xiong'er volcanic belt was most likely a Mesoproterozoic continental magmatic arc that developed along the southern margin of the NCC. This indicates that like North America, Greenland, Baltica, Amazonia, and North Australia, the NCC also underwent subduction-related, continental margin accretion on its southern margin during the outgrowth of the Columbia supercontinent during Paleo-Mesoproterozoic time. This is important for positioning the NCC in the Columbia supercontinent, that is, in any configurations of the Columbia supercontinent, the southern margin of the NCC must be faced an open ocean, not connected to some other continent as proposed by Hou et al. (2008b).

5.3 MESO-NEOPROTEROZOIC EXTENSION AND RIFTING OF THE NCC

Available data show that although the southern margin of the NCC underwent accretion during Mesoproterozoic time, forming the

Xiong'er volcanic belt, its interior experienced a long-lived extension during Meso-Neoproterozoic time, forming the 1.78−1.75 Ga mafic dyke swarms in the central part of the NCC, 1.75−1.68 Ga anorthosite−mangerite−alkali granite−rapakivi granite suites (AMGRS) in the northern part of the craton, and a superlarge Meso-Neoproterozoic fault-controlled basin in the central part of the NCC, called the Yan-Liao Aulacogen in which the Meoproterozoic Changcheng and Jixian groups and the Neoproterozoic Qingbaikou Group were formed (Lu et al., 2008b). Meanwhile, the craton experienced widespread rifting along its northern margin.

5.3.1 1.78−1.75 Ga Mafic Dyke Swarms and 1.75−1.68 Ga AMGRS

Shortly after its final amalgamation at ∼1.85 Ga, the NCC underwent extension inducing the emplacement of 1.78−1.75 Ga mafic dyke swarms in its central part and 1.75−1.68 Ga AMGRS in its northern part. Spatially, the 1.78−1.75 Ga mafic dyke swarms are largely distributed along the TNCO, but cover a larger area than the latter. These mafic dykes are especially widespread in the Hengshan−Wutai−Fuping area of the TNCO, with a predominant NW−SE to NNW−SSE trend and, on a regional scale, constitute dike swarms (Figure 5.6). Generally, they dip steeply, cutting both the Archean and Paleoproterozoic basement rocks and are covered by the Changcheng and Jixian groups. Individual dykes range in width from 10 to 50 m to a maximum of about 100 m, and in length from 10 to 40 km, to a maximum of about 100 km. These dykes are generally unmetamorphosed and undeformed, with chilled contacts. Halls et al. (2000) obtained an upper concordia intercept zircon age of 1769.1 ± 2.5 Ma for a diabase dike in the Hengshan Complex (Table 4.2). Applying the same technique, Peng (2005) obtained a similar upper intercept age of 1754 ± 71 Ma for a mafic dyke in the Wutai Complex (Table 4.2). Wang et al. (2004b) obtained whole-rock $^{40}Ar−^{39}Ar$ ages of 1765.3 ± 1.1, 1774.7 ± 0.7, and 1780.7 ± 0.5 Ma for mafic dykes in the Zanhuang Complex (Table 4.2). All these data suggest that emplacement of mafic dykes in the TNCO occurred in the period 1780−1750 Ma.

The AMGRS in the NCC is mainly distributed in the northern sector of the TNCO. As a main member of the suite, anorthosites are mainly exposed in the Damiao area of the Chengde Complex, extending 40 km long and 10 km wide, with a single outcrop up to nearly

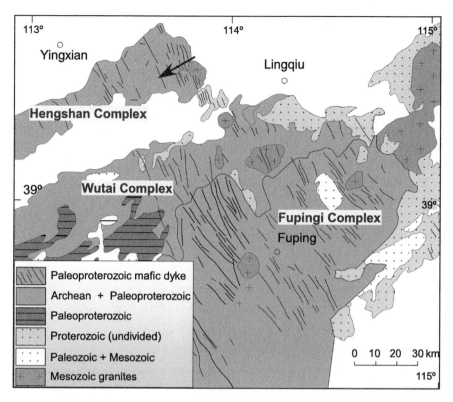

Figure 5.6 Spatial distribution of post-tectonic mafic dikes in the Hengshan−Wutai−Fuping Belt. After Zhao et al. (2001a).

88 km^2. The anorthosites and associated norites and gabbro-norites intrude the Neoarchean gneisses and are covered by the Mesozoic volcanic-sedimentary beds. In the Damiao anorthosites, there are high tenor V−Ti magnetite ore deposits that were formed by residual magmatic segregation that injected the anorthosites, and are spatially associated with the norites. Magmatic zircons from two samples collected from the Damiao anorthosites yielded single-grain U−Pb zircon ages of 1693 ± 7 and 1715 ± 6 Ma (Zhao et al., 2004c). Associated with the anorthosites are gabbros and mangerites of which the former is composed mainly of dark plagioclase (An = 50∼65%, 40∼75% in total minerals), hypersthene (20∼45%), clinopyroxene (Ti-bearing augite 10% or less), and disseminated magnetoilmenite and apatite, whereas the latter intrudes the upper part of the Damiao anorthosite body and consists of microperthite (30%), plagioclase (35%), orthopyroxene and clinopyroxene (30%), and accessory minerals of apatite and Fe−Ti

oxides (5%), with a medium- to coarse-grained or porphyritic texture (Lu et al., 2008b). Another important member in the AMGRS is rapakivi granite that mainly crop out along the east-west Miyun−Qiangzilu−Xinglong Fault and are characterized by megacrysts of alkali feldspars mantled by plagioclase, with a petrographic feature similar to classic rapakivi granites in the world (Zhang et al., 2007b; Lu et al., 2008b). Three rapakivi-granite samples collected from the Shachang body of Miyun County yielded U−Pb zircon ages of 1716 ± 31 (Lu et al., 2008b), 1681 ± 10, and 1679 ± 10 Ma (Yang et al., 2005). It is noteworthy that all four members of the AMGRS (anorthosite, mangerite, alkali-granite, and rapakivi) are exposed along the Chicheng Deep Fault in the northern part of the TNCO, including the Xindi anorthosite, Lanyin quartz syenite, Gudongkou K-rich granite, Gubeikou K-rich granite, Chicheng rapakivi granite, and Chicheng K-rich granite, of which the Lanyin quartz syenite, Chicheng rapakivi granite and Chicheng K-rich granite yielded U−Pb ages of 1697 ± 1, 1702 ± 104, and 1696.7 ± 1.8 Ma, respectively (Yu et al., 1996).

Most researchers argue that the 1.78−1.75 Ga mafic dyke swarms in the central part of the NCC and 1.75−1.68 Ga AMGRS in the northern part of the craton may have resulted from the fragmentation and breakup of the Columbia supercontinent. For example, Hou et al. (2008b) and Peng et al. (2008) suggested that these anorogenic igneous assemblages, together with the Xiong'er volcanic rocks, constitute a LIP formed by a supermantle plume that led to the breakup of the Columbia supercontinent. However, Zhang et al. (2007b) demonstrated that all components of the AMGRS in the northern part of the NCC possess high La/Yb (>10), intermediate Th/Ta ratios, $\varepsilon Nd(T)$ values of -5.0 to -6.3 (with T_{DM} model ages of 2.34−2.58 Ga), and zircon $\varepsilon Hf(T)$ values ranging from -4.1 to -7.5 (with Hf isotopic model T_{DM} ranging from 2.32 to 2.43 Ga), suggesting that the parental magmas of the AMGRS could have been derived from the I-type enriched mantle (EM-I) lithosphere mantle enriched by recycled Archean continental materials, accompanied by some assimilation of lower crustal components. These geochemical features have led Zhang et al. (2007b) to conclude that the 1.75−1.68 Ga AMGRS, together with the 1.78−1.75 Ga mafic dyke swarms in the NCC, were not the products of a mantle plume event that led to the breakup of the Columbia supercontinent, but resulted from a postcollisional extensional event following the amalgamation of the Western and Eastern blocks at ~1.85 Ga. This is supported by Zhao et al. (2009b) who, on the basis of geochemical and Nd–Hf isotope data, suggested that the Damiao anorthosites in the northern sector of the TNCO were derived from extensive melting of lower crustal rocks dragged into the mantle during collision of the Eastern and Western blocks at ~1.85 Ga. Such a model is consistent with the spatial extension of these anorogenic magmatic suites, of which the 1.78–1.75 Ga mafic dyke swarms mainly crop out within the TNCO and the 1.75–1.68 Ga AMGRS are restricted to the northern segment of the TNCO (Zhao et al., 2001a; Zhang et al., 2007b, 2009b). Moreover, the 1.78–1.75 Ga mafic dyke swarms and 1.75–1.68 Ga AMGRS in the NCC are older than those 1.6–1.2 Ga anorthosite–mangerite–charnockite–rapakivi (AMCR) suites and 1.4–1.2 Ga mafic dyke swarms in other cratonic blocks that are considered to have resulted from the fragmentation and breakup of the Columbia supercontinent (Zhao et al., 2002a, 2004a). For these reasons, the 1.78–1.75 Ga mafic dyke swarms and 1.75–1.68 Ga AMGRS in the NCC were most likely related to postcollisional extension following collision between the Western and Eastern blocks at ~1.85 Ga, whereas the geological records of fragmentation and breakup of the Columbia supercontinent in the NCC are the 1.65–0.8 Ga Yan-Liao Aulacogen in the central part of the NCC and the 1.6–1.2 Ga Zhaertai–Bayan Obo rift zone and 1.35 Ga mafic sills at the northern margin of the NCC (Zhao et al., 2011a).

5.3.2 Meso-Neoproterozoic Changcheng, Jixian, and Qingbaikou Groups in the Yan-Liao Aulacogen

The Yan-Liao Aulacogen is a long-lived fault-controlled basin where the deposition of thick sedimentary strata lasted from 1.65 to 0.80 Ga (Li et al., 2011c; Zhang et al., 2013). The strata have been divided into Mesoproterozoic Changcheng and Jixian groups and early Neoproterozoic Qingbaikou Group, with a total thickness of ~9000 m (Figure 5.7). The lowermost Changcheng Group consists of the Changzhougou, Chuanlinggou, Tuanshanzi, and Dahongyu formations from the lower to upper, of which the Changzhougou Formation unconformably overlies the early Precambrian granitic gneiss and comprises conglomerate, pebble-bearing sandstone and arkose sandstone of fluvial facies in the lower portion and sandstone of marine facies in the middle-upper parts. The formation is conformably overlain by the Chuanlinggou Formation that is composed mostly of shales, and is in turn conformably overlain by the dolomite-dominated Tuanshanzi

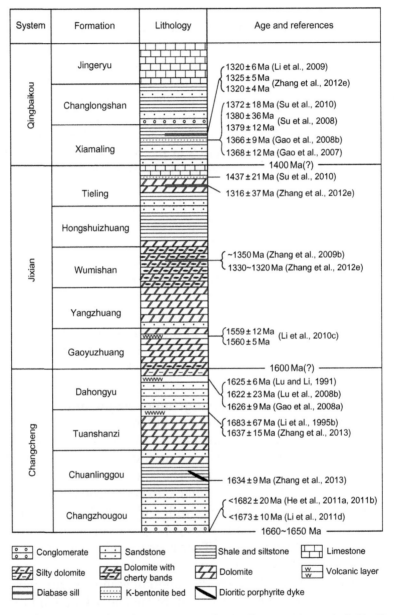

Figure 5.7 Subdivision and zircon U−Pb age constraints on the Meso-Neoproterozoic strata in the Yan-Liao area in the NCC (Zhang et al., 2013).

Formation. Conformably overlying the Tuanshanzi Formation is the uppermost Dahongyu Formation that consists of littoral and neritic sandstone, shale, and potassium-rich trachyte with cherty dolostone in the upper part (Lu et al., 2008b). Traditionally, the Changcheng Group was considered to have deposited in the period 1.8−1.6 Ga (Wan et al., 2003), but recently Li et al. (2011d) obtained a SHRIMP U−Pb zircon age of 1673 ± 10 Ma for a granite dyke that was overlain by the Changzhougou Formation, demonstrating that the Changcheng Group must have been deposited at some time after 1673 ± 10 Ma (Figure 5.7). Most recently, Zhang et al. (2013) obtained SHRIMP U−Pb zircon ages of 1634 ± 9 Ma for a dioritic porphyry dyke intruding the Chuanlinggou Formation and 1637 ± 15 Ma for a potassium-rich volcanic rock within the Tuanshanzi Formation (Figure 5.7), indicating that the Chuanlinggou Formation must have formed at some time before 1634 ± 9 Ma and the Tuanshanzi Formation formed about 1637 ± 15 Ma ago. Volcanic rocks (bentonite) interlayered in the Dahongyu Formation yield zircon ages of 1626 ± 8 Ma (Gao et al., 2008b; Figure 5.7), 1622 ± 23 (Lu et al., 2008b), and 1625 ± 6 Ma (Lu and Li, 1991), confirming the youngest depositional time of the Changcheng Group at ∼ 1.6 Ga.

The Mesoproterozoic Jixian Group consists of four conformable formations with a total thickness of about 4500 m. The lowermost Yangzhuang Formation is made up dominantly of marine alternating red and white dolomite with minor limestone. The overlying Wumishan Formation is about 3300 m thick, the thickest unit in the Jixian Group, and is characterized by an organism-enriching sedimentary cycle that includes gray thin bedded sand-bearing argillaceous dolomicrites (unit A), brown thin bedded and undulate layered microbial dolostones (unit B), dark thick bedded or massive thrombolite and conical stromatolite (unit C), medium or thin bedded algal mat dolostones (unit D), and thin bedded dolostone with white siliceous crust (unit E). Conformably overlying the Wumishan Formation is the Hongshuizhuang Formation that consists predominantly of black shales, only 131 m thick. The uppermost Tieling Formation consists of shale in the lower sequence and limestone in the upper sequence, with a total thickness of 330 m.

The Neoproterozoic Qingbaikou Group was used to be divided into three formations: Xiamaling, Changlongshan, and Jingeryu (Figure 5.7), of which the basal Xiamaling Formation consists of conglomerate, sandstone, shale, and minor bentonites, which are locally conformably overlain by the Changlongshan Formation (siltstone and shale) but in many places are directly covered by the Jingeryu Formation that comprises conglomerate, sandstone, glauconitic sandstone, shale, and limestone. Recently, however, researchers have obtained a large amount of SHRIMP U–Pb zircon ages of ~1.36 Ga for the bentonites of the Xiamaling Formation (Figure 5.7; Gao et al., 2007, 2008a,b; Su et al., 2008, 2010; Li et al., 2009; Zhang et al., 2012e), indicating that the formation was formed in the Mesoproterozoic, not in the Neoproterozoic as conventionally considered. Therefore, the Xiamaling Formation should be removed out from the Qingbaikou Group and assigned to the Jixian Group. The rest two formations in the Qingbaikou Group are still considered to have formed during Neoproterozoic time, which is supported by a large number of glauconite K–Ar ages obtained for the Jingeryu Formation, most of which range from 900 to 800 Ma (Lu, 1992).

In most areas in the NCC, the late Neoproterozoic (800–542 Ma) strata are absent, and the Jingeryu Formation of the Qingbaikou Group is directly disconformably overlain by the middle sequence of the lower Cambrian Fujunshan Formation, in which trilobites of *redlichia chinensis* (Walcott) and *megapalaeolenus fengyangensis* (Chu) were discovered. Only in the Xuhuai and Liaonan areas of the Eastern Block can be seen minor late Neoproterozoic strata named the Yongnian, Xihe, Wuhangshan, and Jinxian formations from lower to upper (Chang, 1980), of which the basal Yongnian Formation consists of conglomerate, pebble-bearing arkose, course arkose, and sandstone. Disconformably overlying the Yongnian Group is the Xihe Formation that comprises conglomerate, glauconite sandstone, siltstone, and shale, which are overlain by the Wuhangshan Formation that consists of shale, siltstone limestone, and dolostone with stromatolites. The uppermost Jinxian Formation consists predominantly of carbonates with abundant stromatolites in the lower and shale, siltstone, and sandstone in the upper. Due to the absence of fossils and reliable isotopic data, the precise depositional ages of these formations remains unknown or controversial, with some researchers proposing that the Yongnian and Xihe formations are the equivalents of the Qingbaikou Group, whereas Wuhangshan and Jinxian formations are interpreted as the equivalents of the Nanhua and Sinian groups in South China.

In summary, the Yan-Liao Aulacogen in the NCC may be one of the most long-lived extensional basins in Earth's history, with deposition lasting for nearly 1.0 billion years. It started to receive deposition about 1.65 Ga ago, largely coincident with global-scale Mesoproterozoic rifting that led to the fragmentation of the Columbia supercontinent, forming a large number of ensialic and continental margin rift sequences, representatives of which are the Belt–Purcell Supergroup in North America, Telemark Supergroup in Baltica, Riphean Aulacogens in Siberia, Kalahari Copper Belt in Southern Africa, and Kibaran Belt in eastern and central Africa. In this sense, like most other cratonic blocks in the world, the NCC was also involved in the Mesoproterozoic fragmentation of the Columbia super-continent, though the Yan-Liao Aulacogen (basin) in the NCC is different from other Mesoproterozoic rifting basins in that the former was a long-lived ensialic basin with deposition from 1.65 to 0.80 Ga.

5.3.3 1.6−1.2 Ga Zhaertai−Bayan Obo Rift Zone

The 1.6−1.2 Ga Zhaertai−Bayan Obo rift zone is exposed at the northern margin of the NCC, extending from Zhaertai in the west, through Bayan Obo and Huade, to Weichang in the east, up to 1500 km long (Figure 5.8; Zhao et al., 2003, 2011a). The best-studied segment of this rift zone is in Bayan Obo, which hosts the largest REE deposit in the world. The zone comprises sedimentary and volcanic assemblages of which the volcanic rocks are typically bimodal in composition, with tholeiitic to alkalic basalts dominant and subordinate K-enriched rhyolites in association with alkaline syenite, aegirinite,

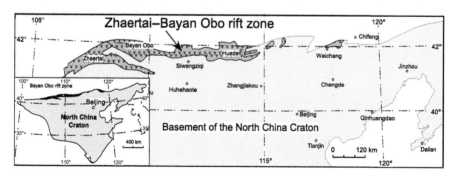

Figure 5.8 Distribution of the late Mesoproterozoic Zhaertai−Bayan Obo rift zone along the northern margin of the NCC (Zhao et al., 2003).

and igneous carbonatites. The sedimentary sequences, conventionally called the "Bayan Obo Group," are composed of conglomerates, pebble-bearing sandstones, turbiditic graywackes, shales, limestones, and iron formations, indicative of a transition from a rapidly subsiding ensialic basin or continental marginal environment to a stable continental shelf environment (Zhou et al., 2002). Available isotope dates show that the rifting magmatism in Bayan Obo clustered in two periods of 1656−1500 and 1312−1223 Ma (Le Bas, 2006 and reference wherein). The younger event was largely coeval with the emplacement of ~1.35 Ga diabase sills recently recognized in the northern margin of the NCC (Zhang et al., 2009b). This further supports the Zhao et al. (2003) interpretation that the 1.6−1.2 Ga Zhaertai−Bayan Obo rift zone on the northern margin of the NCC developed as a result of the fragmentation and breakup of the Paleo-Mesoproterozoic Columbia supercontinent. Thus, in any configurations of Columbia, the northern margin of the NCC should be connected to some other continental block (Zhao et al., 2003, 2011a). In their initial configuration of the Columbia supercontinent, Zhao et al. (2002a, 2004a) proposed a North China−India link where the northern margin of the NCC is placed adjacent to the western margin of the Indian Block, with the TNCO representing the continuation of the Central Indian Tectonic Zone. Such a connection may have maintained until the final breakup of the Columbia supercontinent by a supermantle plume event at ~1.3 Ga, which led to the breakup of the North China−India Block, forming the Zhaertai−Bayan Obo rift zone on the northern margin of the NCC. This Archean-Paleo-Mesoproterozoic North China−India connection has recently been supported by new paleomagnetic data of Zhang et al. (2012f).

REFERENCES

Bai, J., 1993. The Precambrian Geology and Pb−Zn Mineralization in the Northern Margin of North China Platform. Geological Publishing House, Beijing.

Bai, J., Dai, F.Y., 1998. Archean crust of China. In: Ma, X.Y., Bai, J. (Eds.), Precambrian Crust Evolution of China. Springer−Geological Publishing House, Beijing, pp. 15−86.

Bai, J., Wang, R.Z., Guo, J.J., 1992. The Major Geologic Events of Early Precambrian and Their Dating in Wutaishan Region. Geological Publishing House, Beijing.

Bai, Y.L., 1980. On the ancient folds in the eastern Hebei Province. Geol. Res. 3, 68−90.

Bedard, J.H., 2006. A catalytic delamination-driven model for coupled genesis of Archaean crust and sub-continental lithospheric mantle. Geochim. Cosmochim. Acta 70, 1188−1214.

Bohlen, S.R., 1991. On the formation of granulites. J. Metamorphic Geol. 9, 223−229.

Bohlen, S.R., Essene, E.J., 1979. A critical evolution of two pyroxene thermometry in Adirondack granulites. Lithos 12, 335−345.

Brown, M., 2006. A duality of thermal regimes is the distinctive characteristic of plate tectonics since the Neoarchean. Geology 34, 961−964.

Brown, M., 2007. Metamorphic conditions in orogenic belts: A record of secular change. Int. Geol. Rev. 49, 193−234.

Brown, M., 2008. Characteristic thermal regimes of plate tectonics and their metamorphic imprint throughout Earth history. In: Condie, K., Pease, V. (Eds.), When Did Plate Tectonics Begin? Geological Society of America Special Paper, 440, pp. 97−128.

Campbell, I.H., 2005. Large Igneous Provinces and the Mantle Plume Hypothesis. Elements 1, 265−269.

Campbell, I.H., Griffiths, R.W., Hill, R.I., 1989. Melting in an Archaean mantle plumes: heads it's basalts, tail its komatiites. Nature 339, 697−699.

Cao, G.Q., 1996. Early Precambrian Geology of Western Shandong. Geological Publishing House, Beijing, pp. 1−210.

Castonguay, S., Tremblay, A., 2003. Tectonic evolution and significance of Silurian−Early Devonian hinterland-directed deformation in the internal Humber zone of the southern Quebec Appalachians. Can. J. Earth Sci. 40, 255−268.

Chang, L.J., Wang, C.Y., Ding, Z.F., 2012. Upper mantle anisotropy beneath North China from shear wave splitting measurements. Tectonophysics 522, 235−242.

Chang, S., 1980. Subdivision and correlation of Late Precambrian in southern Liaodong Peninsula. In: Wang, Y.L. (Ed.), Sinian Suberathem in China. Tianjin Science and Technology Press, Tianjin, pp. 266−287.

Chen, G.Y., 1983. Komatiites in the second mining area of the Gongchangling region. Chengdu J. Earth Sci. 10, 20−25 (in Chinese with English abstract).

Chen, L., 2007. Geochemistry and geochronology of the Guyang greenstone belt. Postdoctoral Report, Institute of Geology and Geophysics, Chinese Academy of Sciences, Beijing.

Chen, L., Wang, T., Zhao, L., Zheng, T., 2008. Distinct lateral variation of lithospheric thickness in the Northeastern North China Craton. Earth Planet. Sci. Lett. 267, 56−68.

Chen, L., Cheng, C., Wei, Z., 2009. Seismic evidence for significant lateral variations in lithospheric thickness beneath the central and western North China Craton. Earth Planet. Sci. Lett. 286, 171−183.

Chen, N.S., Wang, R.J., Shan, W.Y., Zhong, Z.Q., 1994. Isobaric cooling P−T−t path of the western section of the Miyun Complex and its tectonic implications. Sci. Geol. Sin. 29, 354−364 (in Chinese).

Chen, Y.J., Fu, S.G., 1992. Gold Mineralization in West Henan. Seismological Press, Beijing, 234 pp.

Chen, Y.J., Fu, S.G., Qiang, L.Z., 1992. Tectonic setting of the Xiong'er and Xiyanghe Groups. Geol. Rev. 38, 325−333.

Chen, Y.J., Peng, Y.J., Liu, Y.W., Grant, G., 2006. Progress in the study of Chronostratigraphy of the "Qinghezhen Group". Geol. Rev. 52, 170−177.

Cheng, C., Chen, L., Yao, H.J., Jiang, M.M., Wang, B.Y., 2012. Distinct variations of crustal shear wave velocity structure and radial anisotropy beneath the North China Craton and tectonic implications. Gondwana Res. 23, 25−38.

Cheng, S.H., Kusky, T.M., 2007. Komatiites from west Shandong, North China craton: implications for plume tectonics. Gondwana Res. 12, 77−83.

Cheng, Y.Q., Xu, H.F., 1991. Komatiites from the late Archean Yanlingguan Formation, Shandong: some new findings. China Geol. 135, 31−32.

Condie, K.C., Kröener, A., 2008. When did plate tectonics begin? Evidence from the geologic record. In: Condie, K.C., Pease, V. (Eds.), When Did Plate Tectonics Begin on Planet Earth. Geological Society of America, pp. 281−294., Special Papers 440.

Condie, K.C., Boryta, M.D., Liu, J.Z., Qian, X.L., 1992. The origin of Khondalites—geochemical evidence from the Archean to Early Proterozoic Granulite Belt in the North China Craton. Precambrian Res. 59, 207−223.

Cui, P.L., Sun, J.G., Sha, D.M., Wang, X.J., Zhang, P., Gu, A.L., et al., 2013. Oldest zircon xenocryst (4.17 Ga) from the North China Craton. Int. Geol. 55, 1902−1908.

Cui, W.Y., Wang, C.Q., Wang, S.G., 1991. Geochemistry and metamorphic P−T−t path of the Jianping Complex in the western Liaoning Province. Acta Petrol. Sin. 7, 13−26 (in Chinese).

Dai, X.Y., Liu, J.H., Shao, J.B., Jiang, R.Q., 1990. Geological features of a granite−greenstone belt in the Jiapigou area, Jilin Province. Precambrian Geol. 4, 51−58 (in Chinese).

Dai, Y.P., Zhang, L.C., Wang, C.L., Liu, L., Cui, M.L., Zhu, M.T., et al., 2012. Genetic type, formation age and tectonic setting of the Waitoushan banded iron formation, Benxi, Liaoning Province. Acta Petrol. Sin. 28, 3574−3594.

Dan, W., Li, X.H., Guo, J.H., Liu, Y., Wang, X.C., 2012. Integrated in situ zircon U−Pb age and Hf−O isotopes for the Helanshan Khondalites in North China Craton: juvenile crustal materials deposited in active or passive continental margin? Precambrian Res. 222−223, 143−158.

Defant, M.J., Jackson, T.E., Drummond, M.S., De Boer, J.Z., Bellon, H., Feigenson, M.D., et al., 1992. The geochemistry of young volcanism throughout western Panama and southeastern Costa Rica: an overview. J. Geol. Soc. London 149, 569−579.

Deng, H., Kusky, T., Polat, A., Wang, L., Wang, J.P., Wang, S.J., 2013. Geochemistry of neoarchean mafic volcanic rocks and late mafic dikes in the Zanhuang Complex, Central Orogenic Belt, North China Craton: implications for geodynamic setting. Lithos 175−176, 193−212.

Diwu, C.R., Sun, Y., Guo, A.L., Wang, W.L., Liu, X.M., 2011. Crustal growth in the North China Craton at ~2.5 Ga: evidence from in situ zircon U−Pb ages, Hf isotopes and whole-rock geochemistry of the Dengfeng Complex. Gondwana Res. 20, 149−170.

Dong, C.Y., Liu, D.Y., Li, J.J., Wan, Y.S., Zhou, H.Y., Li, C.D., et al., 2007. Paleoproterozoic Khondalite Belt in the western North China Craton: new evidence from SHRIMP dating and Hf isotope composition of zircons from metamorphic rocks in the Bayan Ul-Helan Mountains area. Chin. Sci. Bull. 52, 2984–2994.

Dong, C.Y., Wan, Y.S., Xu, Z.Y., Liu, D.Y., Yang, Z.S., Ma, M.Z., et al., 2013. SHRIMP zircon U–Pb dating of late Paleoproterozoic kondalites in the Daqing Mountains area on the North China Craton. Sci. China-Earth Sci. 56, 115–125.

Dong, X.J., 2012. Composition and Evolution of the Early Precambrian Basement in the Daqingshan Region, Inner Mongolia (Ph.D. thesis). Changchun University of Science and Technology, Changchun (in Chinese).

Dong, X.J., Xu, Z.Y., Liu, Z.H., Sha, Q., 2012a. Zircon U–Pb geochronology of Archean high-grade metamorphic from Xi Ulanbulang area, Central inner Mongolia. Sci. China Earth Sci. 55, 204–212.

Dong, X.J., Xu, Z.Y., Liu, Z.H., Sha, Q., 2012b. Discovery of 2.7 Ga granitic gneiss in the northern Daqingshan area, Inner Mongolia and its geological significance. Earth Sci. J. China Univ. Geosci. 37, 20–27.

Du, L.L., Zhuang, Y.X., Yang, C.H., Wan, Y.S., Wand, X.S., 2003. Characters of zircons in the Mengjiatun Formation in Xintai of Shandong and their chronological significance. Acta Geol. Sin. 77, 359–366.

Du, L.L., Yang, C.H., Zhuang, Y.X., Wei, R.Z., Wan, Y.S., Ren, L.D., et al., 2010a. Hf isotopic compositions of zircons from 2.7 Ga metasedimentary rocks and biotite plagioclase gneiss in the Mengjiatun formation complex, western Shandong Province. Acta Geol. Sin. 84 (7), 991–1001 (in Chinese with English abstract).

Du, L.L., Yang, C.H., Guo, J.H., Wang, W., Ren, L.D., Wan, Y.S., et al., 2010b. The age of the base of the Paleoproterozoic Hutuo Group in the Wutai Mountains area, North China Craton: SHRIMP zircon U–Pb dating of basaltic andesite. Chin. Sci. Bull. 55, 1782–1789.

Ellis, D., Green, D.H., 1979. An experimental study of the effect of Ca upon garnet-clinopyroxene Fe–Mg exchange equilibria. Contrib. Mineral. Petrol. 71, 13–22.

Escher, A., Beaumont, C., 1997. Formation, burial and exhumation of basement nappes at crustal scale: a geometric model based on the Western Swiss-Italian Alps. J. Struct. Geol. 19, 955–974.

Faure, M., Lin, W., Monie, P., Bruguier, O., 2004. Palaeoproterozoic arc magmatism and collision in Liaodong Peninsula (north-east China). Terra Nova 16, 75–80.

Faure, M., Trap, P., Lin, W., Monie, P., Bruguier, O., 2007. Polyorogenic evolution of the Paleoproterozoic Trans-North China Belt, new insights from the Lüliangshan-Hengshan-Wutaishan and Fuping massifs. Episodes 30, 1–12.

Ganguly, J., Bhattacharya, S., Chakraborty, S., 1988. Convolution effect in the determination of compositional profiles and diffusion co-efficients by microprobe step scans. Am. Mineral. 73, 901–909.

Gao, L.Z., Zhang, C.H., Shi, X.Y., Zhou, H.R., Wang, Z.Q., 2007. Zircon SHRIMP U–Pb dating of the tuff bed in the Xiamaling Formation of the Qingbaikouan System in North China. Geol. Bull. China 26, 249–255 (in Chinese with English abstract).

Gao, L.Z., Zhang, C.H., Shi, X.Y., Wang, Z.Q., Liu, Y., Liu, P., et al., 2008a. SHRIMP zircon ages: basis for refining the chronostratigraphic classification of the Meso-and Neoproterozoic strata in North China Old Land. Acta Geosci. Sin. 29, 366–376 (in Chinese with English abstract).

Gao, L.Z., Zhang, C.H., Shi, X.Y., Song, B., Wang, Z.Q., Liu, Y.M., 2008b. Mesoproterozoic age for Xiamaling Formation in North China Plate indicated by zircon SHRIMP dating. Chin. Sci. Bull. 53, 2665–2671.

Ge, W.C., Zhao, G.C., Sun, D.Y., Wu, F.Y., Lin, Q., 2003. Metamorphic P–T path of the Southern Jilin complex: implications for tectonic evolution of the Eastern Block of the North China Craton. Int. Geol. Rev. 45, 1029–1043.

Ge, X.H., Liu, J.L., 2000. Broken "Western China Craton". Acta Petrol. Sin. 16, 59–66.

Geng, Y.S., Liu, D.Y., Song, B., 1997. Chronological framework of the early Precambrian important events of the Northwestern Hebei granulite terrain. Acta Geol. Sin. 71, 316–327 (English edition, in Chinese with English abstract).

Geng, Y.S., Shen, Q.H., Zhang, Z.Q., 1999. Sm-Nd isotope age and its geological significance of epimetamorphic rock series in the Qinglong area, eastern Hebei Province. Reg. Geol. China 18, 271–276 (in Chinese with English abstract).

Geng, Y.S., Wan, Y.S., Shen, Q.H., Li, H.M., Zhang, R.X., 2000. Chronological framework of the Early Precambrian important events in the Lüliang Area, Shanxi Province. Acta Geol. Sin. 74, 216–223 (English edition).

Geng, Y.S., Wang, X.S., Shen, Q.H., Wu, C.M., 2002. The discovery of Neoproterozoic Jinningian deformed granites in Alax area and its significance. Acta Petrol. Mineral. 21, 412–420.

Geng, Y.S., Liu, F.L., Yang, C.H., 2006. Magmatic event at the end of the Archean in eastern Hebei Province and its geological implication. Acta Geol. Sin. 80, 819–833 (English edition).

Geng, Y.S., Shen, Q.H., Ren, L.D., 2010. Late Neoarchean to Early Paleoproterozoic magmatic events and tectonothermal systems in the North China Craton. Acta Petrol. Sin. 26, 1945–1966.

Geng, Y.S., Du, D.L., Ren, L.D., 2012. Growth and reworking of the early Precambrian continental crust in the North China Craton: constraints from zircon Hf isotopes. Gondwana Res 21, 517–529.

Goldblatt, C., Zahnle, K.J., Sleep, N.H., Nisbet, E.G., 2010. The Eons of Chaos and Hades. Solid Earth 1, 1–3.

Gong, J.H., Zhang, J.X., Yu, S.Y., Li, H.K., Hou, K.J., 2012. ~2.5 Ga TTG rocks from the Alashan terrane and their geological implications. Chin. Sci. Bull. 57, 2715–2728.

Grant, M.L., Wilde, S.A., Wu, F.Y., Yang, J.H., 2009. The application of zircon cathodoluminescence imaging, Th–U–Pb chemistry and U–Pb ages in interpreting discrete magmatic and high-grade metamorphic events in the North China Craton at the Archean/Proterozoic boundary. Chem. Geol. 261, 155–171.

Guan, H., Sun, M., Wilde, S.A., Zhou, X.H., Zhai, M.G., 2002. SHRIMP U-Pb zircon geochronology of the Fuping Complex: implications for formation and assembly of the North China Craton. Precambrian Res. 113, 1–18.

Guo, J.H., Shi, X., 1996. High-pressure granulites, retrograded eclogites and granites in the Early Precambrian Sanggan structural belt. In: Zhai, M.G. (Ed.), Granulites and Lower Continental Crust in the North China Craton. Seismological Press, Beijing, pp. 21–54.

Guo, J.H., Zhai, M.G., 2001. Sm–Nd age dating of high-pressure granulites and amphibolite from Sanggan area, North China Craton. Chin. Sci. Bull. 46, 106–111.

Guo, J.H., Zhai, M.G., Zhang, Y.G., 1993. Early Precambrian Manjinggou high-pressure granulites melange belt on the southern edge of the Huaian Complex, North China Craton: geological features, petrology and isotopic geochronology. Acta Petrol. Sin. 9, 329–341.

Guo, J.H., Shi, X., Bian, A.G., Xu, R.H., Zhai, M.G., Li, Y.G., 1999. Pb isotopic compositions of feldspar and U–Pb age of zircons from early Proterozoic granites in the Sanggan area, North China Craton: metamorphism, crustal melting and tectonothermal events. Acta Petrol. Sin. 15, 199–207.

Guo, J.H., Wang, S.S., Sang, H.Q., Zhai, M.G., 2001. Ar-40-Ar-39 age spectra of garnet porphyroblast: Implications for metamorphic age of high-pressure granulite in the North China Craton. Acta Petrol. Sin. 17, 436–442.

Guo, J.H., O'Brien, P.J., Zhai, M.G., 2002. High-pressure granulites in the Sangan area, North China Craton: metamorphic evolution, P-T paths and geotectonic significance. J. Metamorphic Geol. 20, 741–756.

Guo, J.H., Sun, M., Zhai, M.G., 2005. Sm–Nd and SHRIMP U–Pb zircon geochronology of high-pressure granulites in the Sanggan area, North China Craton: timing of Paleoproterozoic continental collision. J. Asian Earth Sci. 24, 629–642.

Guo, J.H., Chen, Y., Peng, P., Liu. F., Chen. L., Zhang, L.Q., 2006. Sapphirine granulite from Daqingshan area, Inner Modolia—1.85 Ga ultra-high temperature (UHT) metamorphism. In: Proceedings of the National Conference on Petrology and Geodynamics in China (Nanjing).

Guo, J.H., Peng, P., Chen, Y., Jiao, S.J., Windley, B.F., 2012. UHT sapphirine granulite metamorphism at 1.93-1.92 Ga caused by gabbronorite intrusions: implications for tectonic evolution of the northern margin of the North China Craton. Precambrian Res. 222–223, 124–142.

Guo, R.R., Liu, S.W., Santosh, M., Li, Q.G., Bai, X., Wang, W., 2013. Geochemistry, zircon U–Pb geochronology and Lu–Hf isotopes of metavolcanics from eastern Hebei reveal Neoarchean subduction tectonics in the North China Craton. Gondwana Res. 24, 664–686.

Guo, S.S., Li, S.G., 2009. SHRIMP zircon U–Pb ages for the Paleoproterozoic metamorphic-magmatic events in the southeast margin of the North China Craton. Sci. China 52, 1039–1045 (Series D-Earth Sciences).

Halls, H.C., Li, J.H., Davis, D., Hou, G., Zhang, B.X., Qian, X.L., 2000. A precisely dated Proterozoic paleomagnetic pole from the North China craton, and its relevance to palaeocontinental reconstruction. Geophys. J. Int. 143, 185–203.

Hamilton, W.B., 1998. Archean magmatism and deformation were not products of plate tectonics. Precambrian Res. 91, 143–179.

Han, B.F., Xu, Z., Ren, R., Li, L.L., Yang, J.H., Heng, Y., 2012. Crustal growth and intracrustal recycling in the middle segment of the Trans-North China Orogen, North China Craton: a case study of the Fuping Complex. Geol. Mag. 149, 729–742.

Hao, D.F., Li, S.Z., Zhao, G.C., Sun, M., Han, Z.Z., Zhao, G.T., 2004. Origin and its constraint to tectonic evolution of Paleoproterozoic granitoids in the eastern Liaoning and Jilin province, North China. Acta Petrol. Sin. 20, 1409–1416.

He, G.P., Ye, H.W., 1998. Two type of Early Proterozoic metamorphism in the eastern Liaoning to southern Jilin and their tectonic implication. Acta Petrol. Sin. 14, 152–162.

He, G.P., Lu, L.Z., Yie, H.W., Jin, S.Q., Yie, T.S., 1991. Metamorphic Evolution of the Early Precambrian Basement of the Eastern Hebei and Southeastern Inner Mongolia Areas. Jilin University Press, Changchun, pp. 1–16 (in Chinese).

He, T.X., Lin, Q., Fang, Z.R., Liu, S.W., Zhao, G.C., Liu, C.H., 1992. Petrogenesis of Archean Granitoid Rocks in Eastern Hebei, China. Jilin Science and Technology Press, Changchun, 171 pp.

He, Y.H., Zhao, G.C., Sun, M., 2008. Geochemistry of the Xiyanghe volcanics at the southern margin of the North China Craton: petrogenesis and tectonic implications. Lithos 102, 158–178.

He, Y.H., Zhao, G.C., Sun, M., Xia, X.P., 2009. SHRIMP and LA-ICP-MS zircon geochronology of the Xiong'er volcanic rocks: implications for the Paleo-Mesoproterozoic evolution of the southern margin of the North China Craton. Precambrian Res. 168, 213–222.

He, Y.H., Zhao, G.C., Sun, M., Han, Y., 2010a. Petrogenesis and tectonic setting of volcanic rocks in the Xiaoshan and Waifangshan areas along the southern margin of the North China Craton: constraints from bulk-rock geochemistry and Sr–Nd isotopic composition. Lithos 114, 186–199.

He, Y.H., Zhao, G.C., Sun, M., 2010b. Geochemical and isotopic study of the Xiong'er volcanic rocks at the southern margin of the North China Craton: petrogenesis and tectonic implications. J. Geol. 118, 417–433.

He, Z.J., Niu, B.G., Zhang, X.Y., Zhao, L., Liu, R.Y., 2011a. Discovery of the paleo-weathered mantle of the rapakivi granite covered by the Proterozoic Changzhougou Formation in the Miyun area, Beijing and their detrital zircon dating. Geol. Bull. China 30, 798–802 (in Chinese with English abstract).

He, Z.J., Zhang, X.Y., Niu, B.G., Liu, R.Y., Zhao, L., 2011b. The paleo-weathering mantle of the Proterozoic rapakivi granite in Miyun County, Beijing and the relationship with the Changzhougou Formation of Changchengian System. Earth Sci. Front. 18, 123–130 (in Chinese with English abstract).

Hill, R.H., Campbell, I.H., Compston, W., 1989. Age and origin of granitic rocks in the Kalgoorlie-Norseman region of Western Australia: implications for the origin of Archaean crust. Geochim. Cosmochim. 53, 1259–1275.

Hill, R.I., Campbell, I.R., Davis, G.F., Griffiths, R.W., 1992. Mantle plumes and continental tectonics. Science 256, 186–193.

Hoffman, P.F., 1997. Tectonic genealogy of North America. In: van der Pluijm, B.A., Marshak, S. (Eds.), Earth Structure: an Introduction to Structural Geology and Tectonics. McGraw-Hill, New York, NY, pp. 459–464.

Holland, T., Blundy, J., 1994. Non-ideal interactions in calcic amphiboles and their bearing on amphibole–plagioclase thermometry. Contrib. Mineral. Petrol. 116, 433–447.

Hou, G.T., Wang, C.C., Wang, Y.X., Xiao, F.F., Li, L., 2008a. SHRIMP Zircon U–Pb chronology and tectonic significance of the latest Neoarchean Mengshan Diorite, Western Shandong Province. Geol. J. China Univ. 14, 22–28.

Hou, G.T., Santosh, M., Qian, X.L., Lister, G.S., Li, J.H., 2008b. Configuration of the Late Paleoproterozoic supercontinent Columbia: insights from radiating mafic dyke swarms. Gondwana Res. 14, 355–409.

Huang, J.Q., 1977. The basic outline of China tectonics. Acta Geol. Sin. 52, 117–135 (in Chinese).

Huang, X., Bai, Z., DePaolo, D.J., 1986. Sm–Nd isotope study of early Archaean rocks, Qianan, Hebei Province, China. Geochim. Cosmochim. Acta 50, 625–631.

Huang, X.L., Niu, Y.L., Xu, Y.G., Yang, Q.J., Zhong, J.W., 2010. Geochemistry of TTG and TTG-like gneisses from Lushan-Taihua complex in the southern North China Craton: implications for late Archean crustal accretion. Precambrian Res. 182, 43–56.

Jahn, B.M., 1989. Origin of granulites: geochemical constrains from Archean granulite facies rocks of the Sino-Korean craton, China. In: Vielzeuf, D., Vidal, P.h. (Eds.), Granulites and Crustal Evolution. Kluwer Academic Publishers, Dordrecht, pp. 471–491.

Jahn, B.M., Ernst, W.G., 1990. Late Archean Sm–Nd isochron age for mafic–ultramafic supracrustal amphibolites from the Northern Sino–Korean craton, China. Precambrian Res. 46, 295–306.

Jahn, B.M., Zhang, Z.Q., 1984. Archean granulite gneisses from eastern Hebei Province, China: rare earth geochemistry and tectonic implications. Contrib. Mineral. Petrol. 85, 224–243.

Jahn, B.M., Auvray, B., Cornichet, J., Bai, Y.D., Shen, Q.H., Liu, D.Y., 1987. 3.5 Ga old amphibolites from eastern Hebei Province, China: field occurrence, petrography, Sm–Nd isochron age and REE geochemistry. Precambrian Res. 34, 311–346.

Jahn, B.M., Auvray, B., Shen, Q.H., Liu, D.Y., Zhang, Z.Q., Dong, Y.J., et al., 1988. Archean crustal evolution in China: the Taishan Complex and evidence for Juvenile crustal addition from long-term depleted mantle. Precambrian Res. 38, 381–403.

Jahn, B.M., Liu, D.Y., Wan, Y.S., Song, B., Wu, J.S., 2008. Archean crustal evolution of the Jiaodong Peninsula, China, as revealed by zircon SHRIMP geochronology, elemental and Nd-isotope geochemistry. Am. J. Sci. 308, 232–269.

Jayananda, M., Moyen, J.F., Martin, H., Peucat, J.J., Auvray, B., Mahabaleswar, B., 2000. Late Archean (2550–2520 Ma) juvenile magmatism in the Eastern Dharwar craton, southern India: constraints from geochronology, Nd–Sr isotopes and whole rock geochemistry. Precambrian Res. 99, 225–254.

Jian, P., Zhang, Q., Liu, D.Y., Jin, W.J., Jia, X.Q., Qian, Q., 2005. SHRIMP dating and geological significance of Late Achaean high-Mg diorite (sanukite) and hornblende-granite at Guyang of Inner Mongolia. Acta Petrol. Sin. 21, 151–157 (in Chinese with English abstract).

Jian, P., Kröner, A., Windley, B.F., Zhang, Q., Zhang, W., Zhang, L.Q., 2012. Episodic mantle melting–crustal reworking in the late Neoarchean of the northwestern North China Craton: zircon ages of magmatic and metamorphic rocks from the Yinshan Block. Precambrian Res. 222–223, 230–254.

Jiang, M.M., Ai, Y.S., Chen, L., Yang, Y.J., 2012. Local modification of the lithosphere beneath the central and western North China Craton: 3-D constraints from Rayleigh wave tomography. Gondwana Res. 24, 849–864.

Jiang, N., Guo, J.H., Zhai, M.G., Zhang, S.Q., 2010. ~2.7 Ga crust growth in the North China craton. Precambrian Res. 179, 37–49.

Jiao, S.J., Guo, J.H., 2011. Application of the two-feldspar geothermometer to ultrahigh-temperature (UHT) rocks in the Khondalite belt, North China craton and its implications. Am. Mineral. 96, 250–260.

Jiao, S.J., Guo, J.H., Mao, Q., Zhao, R.F., 2011. Application of Zr-in-rutile thermometry: a case study from ultrahigh-temperature granulites of the Khondalite Belt, North China Craton. Contrib. Mineral. Petrol. 162, 379–393.

Jiao, S.J., Guo, J.H., Harley, S.L., Windley, B.F., 2013a. New constraints from garnetite on the P-T path of the Khondalite Belt: implications for the tectonic evolution of the North China Craton. J. Petrol. 54, 1725–1758.

Jiao, S.J., Guo, J.H., Harley, S.L., Peng, P., 2013b. Geochronology and trace element geochemistry of zircon, monazite and garnet from the garnetite and/or associated other high-grade rocks: implications for Palaeoproterozoic tectonothermal evolution of the Khondalite Belt, North China Craton. Precambrian Res. 237, 78–100.

Jin, W., 1989. Geological Evolution and Metamorphic Dynamics of Early Precambrian Basement Rocks Along the Northern Boundary (Central Section) of the North China Craton (Ph.D. thesis). Changchun University of Science and Technology, Changchun (in Chinese).

Jin, W., Li, S.X., 1994. The lithologies and geological features of the Palaeoproterozoic orogenic belt in Daqingshan area, Inner Mongolia. In: Qian, X.L., Wang, R.M. (Eds.), Geological Evolution of the Granulite Terranes in the North Part of the North China Craton. Seismological Press, Beijing, pp. 32–42. (in Chinese with English abstract).

Jin, W., Li, S.X., 1996. PTt Path and crustal thermodynamic model of Late Archaean-Early Proterozoic high grade metamorphic terrain in North China. Acta Petrol. Sin. 12, 209–221.

Jin, W., Li, S.X., Liu, X.S., 1991. The Metamorphic dynamics of Early Precambrian high-grade metamorphic rocks series in Daqing-Ulashan area, Inner Mongolia. Acta Petrol. Sin. 7, 27–35.

Kepezhinskas, P.K., Defant, M.J., Drummond, M.S., 1996. Progressive enrichment of island arc mantle by melt-peridotite interaction inferred from Kamchatka xenoliths. Geochim. Cosmochim. Acta 60, 1217–1229.

Kröner, A., Compston, W., Zhang, G.W., Guo, A.L., Todt, W., 1988. Ages and tectonic setting of Late Archean greenstone–gneiss terrain in Henan Province, China, as revealed by single-grain zircon dating. Geology 16, 211–215.

Kröner, A., Cui, W.Y., Wang, W.Y., Wang, C.Q., Nemchin, A.A., 1998. Single zircon ages from high-grade rocks of the Jianping Complex, Liaoning Province, NE China. J. Asian Earth Sci. 16, 519–532.

Kröner, A., Wilde, S.A., Li, J.H., Wang, K.Y., 2005a. Age and evolution of a late Archaean to early Palaeozoic upper to lower crustal section in the Wutaishan/Hengshan/Fuping terrain of northern China. J. Asian Earth Sci. 24, 577−595.

Kröner, A., Wilde, S.A., O'Brien, P.J., Li, J.H., Passchier, C.W., Walte, N.P., et al., 2005b. Field relationships, geochemistry, zircon ages and evolution of a late Archean to Paleoproterozoic lower crustal section in the Hengshan Terrain of Northern China. Acta Geol. Sin. 79, 605−629 (English edition).

Kröner, A., Wilde, S.A., Zhao, G.C., O'Brien, P.J., Sun, M., Liu, D.Y., et al., 2006. Zircon geochronology of mafic dykes in the Hengshan Complex of northern China: evidence for late Palaeoproterozoic rifting and subsequent high-pressure event in the North China Craton. Precambrian Res. 146, 45−67.

Kusky, T.M., 2011a. Geophysical and geological tests of tectonic models of the North China Craton. Gondwana Res. 20, 26−35.

Kusky, T.M., 2011b. Comparison of results of recent seismic profiles with tectonic models of the North China craton. J. Earth Sci. 22, 250−259.

Kusky, T.M., Li, J.H., 2003. Paleoproterozoic tectonic evolution of the North China Craton. J. Asian Earth Sci. 22, 383−397.

Kusky, T.M., Santosh, M., 2009. The Columbia Connection in North China. In: Reddy, S.M., Mazumder, R., Evans, D., Collins, A.S. (Eds.), Paleoproterozoic Supercontinents and Global Evolution. Geological Society of London, pp. 49−71., Special Publication 323.

Kusky, T.M., Li, J.H., Tucker, R.D., 2001. The Archean Dongwanzi ophiolite complex, North China Craton: 2. 505-billion-year-old oceanic crust and mantle. Science 292, 1142−1145.

Kusky, T.M., Li, J.H., Santosh, M., 2007. The Paleoproterozoic North Hebei Orogen: North China Craton's collisional suture with the Columbia supercontinent. Gondwana Res. 12, 4−28.

Lai, X.D., Yang, X.Y., 2012. Characteristics of the banded iron formation (BIF) and its zircon chronology in Yangzhuang, western Shandong. Acta Petrol. Sin. 28, 3612−3622.

Le Bas, M.J., 2006. Re-interpretation of zircon date in a carbonatite dyke at the Bayan Obo giant REE−Fe−Nb deposit, China. Acta Petrol. Sin. 22, 517−518.

Li, H.K., Li, H.M., Lu, S.N., 1995b. Grain zircon U−Pb age for volcanic rocks from Tuanshanzi Formation of Changcheng System and their geological implication. Geochimica 24, 43−47 (in Chinese with English abstract).

Li, H.K., Lu, S.N., Li, H.M., Sun, L.X., Xiang, Z.Q., Geng, J.Z., et al., 2009. Zircon and beddeleyite U−Pb precision dating of basic rock sills intruding Xiamaling Formation, North China. Geol. Bull. China 28, 1396−1404.

Li, H.K., Zhu, S.X., Xiang, Z.Q., Su, W.B., Lu, S.N., Zhou, H.Y., et al., 2010c. Zircon U−Pb dating on tuff bed from Gaoyuzhuang Formation in Yanqiang, Beijing: further constraints on the new subdivision of the Mesoproterozoic stratigraphy in the northern North China Craton. Acta Petrol. Sin. 26, 2131−2140 (in Chinese with English abstract).

Li, H.K., Su, W.B., Zhou, H.Y., Geng, J.Z., Xiang, Z.Q., Cui, Y.R., et al., 2011c. The base age of the Changchengian System at the northern North China Craton should be younger than 1670 Ma: constraints from zircon U−Pb LA-ICP-MS dating of a granite-porphyry dyke in Miyun County, Beijing. Earth Sci. Front. 18, 108−120.

Li, J.H., Kusky, T.M., 2007. A late Archean foreland fold and thrust belt in the North China Craton: implications for early collisional tectonics. Gondwana Res. 12, 47−66.

Li, J.H., Liu, W.J., 1996. Late Archean tectonic evolution. In: Zhai, M.G. (Ed.), Granulites and Lower Continental Crust in the North China Archean Craton. Seismological Press, Beijing, pp. 207−239.

Li, J.H., Qian, Q.L., 1991. A study on the Longquanguan shear zone in the northern part of the Taihang Mountains. Shanxi Geol. 6, 17−29 (in Chinese).

Li, J.H., Zhai, M.G., Li, Y.G., Zhan, Y.G., 1998. Discovery of Late Archean high-pressure granulites in Luanping-Chengde area, Northern Hebei Province: tectonic implications. Acta Petrol. Sin. 14, 34−41.

Li, J.H., Qian, X.L., Liu, S.W., 1999. Geochemistry of the khondalite series in the central North China Craton and implications for the crustal cratonization. Sci. China 29, 193−203 (Series D).

Li, J.H., Kröner, A., Qian, X.L., O'Brien, P., 2000. Tectonic evolution of an Early Precambrian high-pressure Granulite Belt in the North China Craton. Acta Geol. Sin. 74, 246−258.

Li, J.J., Shen, B.F., 2000. Geochronology of Precambrian continent crust in Liaoning Province and Jilin Province. Prog. Precambrian Res. 23, 244−249.

Li, J.L., 1982. REE geochemistry of early Archean Taipingzhai greenstones in Eastern Hebei. Chin. Sci. Bull. 19, 1192−1196.

Li, J.L., Wang, K.Y., Wang, C.Q., Liu, X.H., Zhao, Z.Y., 1990. An Early Proterozoic collision belt in the Wutaishan area, China. Sci. Geol. Sin. 25, 1−11 (in Chinese).

Li, J.Y., 2004. Structural characteristics of crustal "mosaicking and superimposition" of the continent of China and its evolution. Geol. Bull. China 23, 986−1004.

Li, S.X., Liu, X.S., Zhang, L.Q., 1987. Granite−greenstone belt in Sheerteng area, Inner Mongolia, China. J. Changchun Coll. Geol. 17, 81−102 (in Chinese).

Li, S.X., Sun, D.Y., Yu, H.F., Jin, W., Liu, X.S., Cao, L., 1995a. Distribution of Ductile Shear Zones and Metallogenic Prediction of the Related Gold Deposits in the Early Precambrian Metamorphic Rocks, Middle-Western Inner Mongolia. Jilin Science and Technology Press, Changchun, pp. 1−111.

Li, S.Z., Zhao, G.C., 2007. SHRIMP U-Pb zircon geochronology of the Liaoji granitoids: constraints on the evolution of the Paleoproterozoic Jiao-Liao-Ji Belt in the Eastern Block of the North China Craton. Precambrian Res. 158, 1−16.

Li, S.Z., Han, Z.Z., Liu, Y.J., Yang, Z.S., Ma, R., 2001. Regional metamorphism of the Liaohe Group: implications for continental dynamics. Geol. Rev. 47, 9−18.

Li, S.Z., Hao, D.F., Zhao, G.C., Sun, M., Han, Z.Z., Guo, X.Y., 2004a. Geochemical features and origin of Dandong granite. Acta Petrol. Sin. 20, 1417−1423.

Li, S.Z., Zhao, G.C., Sun, M., Wu, F.Y., Liu, J.Z., Hao, D.F., et al., 2004b. Mesozoic, not Paleoproterozoic SHRIMP U−Pb zircon ages of two Liaoji granites, Eastern Block, North China Craton. Int. Geol. Rev. 46, 162−176.

Li, S.Z., Zhao, G.C., Sun, M., Wu, F.Y., Hao, D.F., Han, Z.Z., et al., 2005. Deformational history of the Paleoproterozoic Liaohe Group in the Eastern Block of the North China Craton. J. Asian Earth Sci. 24, 654−669.

Li, S.Z., Zhao, G.C., Sun, M., Han, Z.Z., Zhao, G.T., Hao, D.F., 2006. Are the South and North Liaohe Groups of the North China Craton different exotic terranes? Nd isotope constraints. Gondwana Res. 9, 198−208.

Li, S.Z., Zhao, G.C., Zhang, J., Sun, M., Zhang, G.W., Luo, D., 2010a. Deformational history of the Hengshan-Wutai-Fuping belt: implications for the evolution of the Trans-North China Orogen. Gondwana Res. 18, 611−631.

Li, S.Z., Zhao, G.C., Santosh, M., Liu, X., Dai, L.M., 2011b. Palaeoproterozoic tectonothermal evolution and deep crustal processes in the Jiao-Liao-Ji Belt, North China Craton: a review. Geol. J. 46, 525−543.

Li, S.Z., Zhao, G.C., Santosh, M., Liu, X., Lai, L.M., Suo, Y.H., et al., 2012. Structural evolution of the Jiaobei Massif in the southern segment of the Jiao-Liao-Ji Belt, North China Craton. Precambrian Res. 200−203, 59−73.

Li, T.S., Zhai, M.G., Peng, P., Chen, L., Guo, J.H., 2010b. Ca. 2.5 billion year old coeval ultramafic-mafic and syenitic dykes in Eastern Hebei: implications for cratonization of the North China Craton. Precambrian Res. 180, 143−155.

Li, X.P., Yang, Z.Y., Zhao, G.C., Grapes, R., Guo, J.H., 2011a. Geochronology of khondalite-series rocks of the Jining Complex: confirmation of depositional age and tectonometamorphic evolution of the North China craton. Int. Geol. Rev. 53, 1194−1211.

Li, Z.L., 1993. Metamorphic P−T−t path of the Archaean rocks in the eastern Shandong Province and its implications. Shandong Geol. 9, 31−41 (in Chinese).

Liu, C.H., Zhao, G.C., Sun, M., Wu, F.Y., Yang, J.H., Yin, C.Q., et al., 2011c. U−Pb and Hf isotopic study of detrital zircons from the Yejishan Group of the Lüliang Complex: constraints on the timing of collision between the Eastern and Western blocks, North China Craton. Sediment. Geol. 236, 129−140.

Liu, C.H., Zhao, G.C., Sun, M., Zhang, J., Yin, C.Q., Wu, F.Y., et al., 2011d. U−Pb and Hf isotopic study of detrital zircons from the Hutuo Group of the Wutai Complex: constraints on the timing of collision between the Eastern and Western blocks, North China Craton. Gondwana Res. 20, 106−121.

Liu, C.H., Zhao, G.C., Sun, M., Zhang, J., Yin, C.Q., He, Y.H., 2012b. Detrital zircons U−Pb dating, Hf isotope and whole-rock geochemistry from the Songshan Group in the Dengfeng Complex: constraints on the tectonic evolution of the Trans-North China Orogen. Precambrian Res. 192−195, 1−156.

Liu, C.H., Zhao, G.C., Sun, M., Zhang, J., Yin, C.Q., 2012c. U−Pb geochronology and Hf isotope geochemistry of detrital zircons from the Zhongtiao Complex: constraints on the tectonic evolution of the Trans-North China Orogen. Precambrian Res. 222−223, 159−172.

Liu, C.H., Zhao, G.C., Liu, F.L., Sun, M., Zhang, J., Yin, C.Q., 2012e. Zircons U−Pb and Lu−Hf isotopic and whole-rock geochemical constraints on the Gantaohe Group in the Zanhuang Complex: implications for the tectonic evolution of the Trans-North China Orogen. Lithos 146−147, 80−92.

Liu, C.H., Liu, F.L., Zhao, G.C., 2013c. Provenance and tectonic setting of the Jiehekou Group in the Lüliang Complex: constraints from zircon U−Pb age and Hf isotopic studies. Acta Petrol. Sin. 29, 517−553.

Liu, D.Y., Shen, Q.H., Zhang, Z.Q., Jahn, B.M., Auvray, B., 1990. Archean crustal evolution in China: U−Pb geochronology of the Qianxi Complex. Precambrian Res. 48, 223−244.

Liu, D.Y., Nutman, A.P., Compston, W., Wu, J.S., Shen, Q.H., 1992. Remnants of ≥3800 Ma crust in the Chinese part of the Sino − Korean craton. Geology 20, 339−342.

Liu, D.Y., Geng, Y.S., Song, B., 1997b. Late Archean crustal accretion and reworking in north-western Hebei Province: geochronological evidence. Acta Petrol. Sin. 18, 226−232 (in Chinese with English abstract).

Liu, D.Y., Wan, Y.S., Wu, J.S., Wilde, S.A., Zhou, H.Y., Dong, C.Y., et al., 2007, Eoarchean rocks and zircons in the North China Craton. In: van Kranendonk, M., Smithies, R.H., Bennett, V. (Eds), Earth's Oldest Rocks, 15, Amsterdam, The Netherlands, Elsevier Series Developments in Precambrian Geology, pp. 251−273.

Liu, D.Y., Wilde, S.A., Wan, Y.S., Wu, J.S., Zhou, H.Y., Dong, C.Y., et al., 2008a. New U−Pb and Hf isotopic data confirm Anshan as the oldest preserved segment of the North China Craton. Am. J. Sci. 308, 200−231.

Liu, D.Y., Wilde, S.W., Wan, Y.S., Wang, S.Y., Valley, J.W., Kita, N., et al., 2009a. Combined U−Pb, hafnium and oxygen isotope analysis of zircons from meta-igneous rocks in the southern North China Craton reveal multiple events in the Late Mesoarchean−Early Neoarchean. Chem. Geol. 261, 140−154.

Liu, F., Guo, J.H., Lu, X.P., Diwu, C.R., 2009b. Crustal growth at ~2.5 Ga in the North China Craton: evidence from whole-rock Nd and zircon Hf isotopes in the Huai'an gneiss terrane. Chin. Sci. Bull. 54, 4704−4713.

Liu, F., Guo, J.H., Peng, P., Qian, Q., 2012a. Zircon U−Pb ages and petrochemistry of the Huai'an TTG gneisses terrane: petrogenesis and implications for ∼2.5 Ga crustal growth in the North China Craton. Precambrian Res. 212−213, 225−244.

Liu, F.L., 1995. Metamorphic Mineral-Fluid Evolution and Tectonic Environments of the Granulite Facies Terrane in the Huaian−Datong Area (Ph.D. dissertation). Changchun University of Science and Technology, Changchun.

Liu, F.L., 1996. Mineral evolution and the significance of garnet of the khondalite series in the area of Hongsipu-Huangtuyao. J. Changchun Earth Sci. 26, 278−284.

Liu, F.L., 1997. The metamorphic reaction and water activity of basic granulite in the Datong-Huaian region. Acta Petrol. Sin. 13, 27−43.

Liu, F.L., Shen, Q.H., 1999. Retrogressive textures and metamorphic reaction features of Al-rich gneisses in the granulite facies belt from northwestern Hebei province. Acta Petrol. Sin. 15, 505−517.

Liu, F.L., Shen, Q.H., Geng, Y.S., Xu, X.C., Ma, R., 1997a. Experimental study of melting reaction and genetic mechanism of mineral phase transformation in granulite facies metamorphism. Acta Geol. Sin. 71, 254−265.

Liu, J.H., 2001. Discovery of komatiites in the Helong area (southern Jilin) in the eastern part of the North China Craton. Geol. Rev. 47, 420−425.

Liu, J.H., Liu, F.L., Ding, Z.J., Liu, C.H., Yang, H., Liu, P.H., et al., 2013b. The growth, reworking and metamorphism of early Precambrian crust in the Jiaobei terrane, the North China Craton: constraints from U−Th−Pb and Lu−Hf isotopic systematics, and REE concentrations of zircon from Archean granitoid gneisses. Precambrian Res. 224, 287−303.

Liu, J.Z., Qiang, X.K., Liu, X.S., Ouyang, Z.Y., 2000a. Dynamics and genetic grids of sapphirine-bearing spinel gneiss in Daqing Mountain orogen zone, Inner Mongolia. Acta Petrol. Sin. 16, 245−255.

Liu, L., Zhang, L.C., Dai, Y.P., Wang, C.L., Li, Z.Q., 2012f. Formation age, geochemical signatures and geological significance of the Sanheming BIF-type iron deposit in the Guyang greenstone belt, Inner Mongolia. Acta Petrol. Sin. 28, 3623−3637.

Liu, S.J., Li, J.H., 2009. Paleoproterozoic high temperature paired Metamorphic Belt in central part of Southern Inner Mongolia and its tectonic implication. Geol. J. China Univ. 15, 48−56.

Liu, S.J., Li, J.H., Santosh, M., 2008b. Ultrahigh temperature metamorphism of Tuguiwula khondalite belt, Inner Mongolia: metamorphic reaction texture and P−T indication. Acta Petrol. Sin. 24, 1185−1192.

Liu, S.J., Li, J.H., Santosh, M., 2010. First application of the revised Ti-in-zircon geothermometer to Paleoproterozoic ultrahigh-temperature granulites of Tuguiwula, Inner Mongolia, North China Craton. Contrib. Mineral. Petrol. 159, 225−235.

Liu, S.J., Wan, Y.S., Sun, H.Y., Nutman, A.P., Xie, H.Q., Dong, C.Y., et al., 2013a. Paleo- to Eoarchean crustal evolution in eastern Hebei, North China Craton: new evidence from SHRIMP U−Pb dating and in-situ Hf isotopic study of detrital zircons from paragneisses. J. Asian Earth Sci. 78, 4−17.

Liu, S.W., 1991. P−T−t paths determined from zonational hornblendes from high-grade metamorphic rocks in Eastern Hebei. J. Changchun Univ. Earth Sci. 21, 151−162 (in Chinese).

Liu, S.W., Liang, H.H., Zhao, G.C., Hua, Y.G., Jian, A.H., 2000b. Isotopic chronology and geological events of Precambrian Complex in the Taihangshan region. Sci. China 43, 386−393 (Series D Earth Sciences).

Liu, S.W., Pan, P.M., Li, J.H., Li, Q.G., Zhang, J., 2002a. Geological and isotopic geochemical constraints on the evolution of the Fuping Complex, North China Craton. Precambrian Res. 117, 41−56.

Liu, S.W., Li, J.H., Pan, Y.M., Zhang, J., Li, Q.G., 2002b. An Archean continental block in the Taihangshan and Hengshan regions: constraints from geochronology and geochemistry. Prog. Nat. Sci. 12, 568−576.

Liu, S.W., Pan, P.M., Xie, Q.L., Zhang, J., Li, Q.G., 2004a. Archean geodynamics in the Central Zone, North China Craton: constraints from geochemistry of two contrasting series of granitoids in the Fuping and Wutai complexes. Precambrian Res. 130, 229–249.

Liu, S.W., Shu, G.M., Pan, Y.M., Dang, Q.N., 2004b. Electron microprobe dating and metamorphic age of Wutai Group, Wutai Mountains. Geol. J. Chinese Univ. 10, 356–363 (in Chinese with English abstract).

Liu, S.W., Pan, P.M., Xie, Q.L., Zhang, J., Li, Q.G., 2005. Geochemistry of the Paleoproterozoic Nanying granitoid gneisses: constraints on the tectonic setting of the Central Zone, North China Craton. J. Asian Earth Sci. 24, 643–658.

Liu, S.W., Zhao, G.C., Wilde, S.A., Shu, G.M., Sun, M., Li, Q.G., et al., 2006. Th–U–Pb monazite geochronology of the Lüliang and Wutai complexes: constraints on the tectonothermal evolution of the Trans-North China Orogen. Precambrian Res. 148, 205–225.

Liu, S.W., Li, Q.G., Liu, C.H., Lü, Y.J., Zhang, F., 2009c. Guandishan granitoids of the Paleoproterozoic Lüliang Metamorphic Complex in the Trans-North China Orogen: SHRIMP zircon ages, petrogenesis and tectonic implications. Acta Geol. Sin. 83, 580–602 (English edition).

Liu, S.W., Santosh, M., Wang, W., Bai, X., Yang, P.T., 2011a. Zircon U–Pb chronology of the Jianping Complex: implications for the Precambrian crustal evolution history of the northern margin of North China Craton. Gondwana Res. 20, 48–63.

Liu, S.W., Lü, Y.J., Wang, W., Yang, P.T., Bai, X., Feng, Y.G., 2011b. Petrogenesis of the Neoarchean granitoid gneisses in northern Hebei Province. Acta Petrol. Sin. 27, 909–921 (in Chinese with English abstract).

Liu, S.W., Zhang, J., Li, Q.G., Zhang, L.F., Wang, W., Yang, P.T., 2012d. Geochemistry and U–Pb zircon ages of metamorphic volcanic rocks of the Paleoproterozoic Lüliang Complex and constraints on the evolution of the Trans-North China Orogen, North China Craton. Precambrian Res. 222–223, 173–190.

Liu, X.S., Jin, W., Li, S.X., Xu, X.C., 1993. Two types of Precambrian high-grade metamorphism, Inner Mongolia, China. J. Metamorphic Geol. 11, 499–510.

Lü, B., Zhai, M.G., Li, T.S., Peng, P., 2012. Zircon U–Pb ages and geochemistry of the Qinglong volcano-sedimentary rock series in Eastern Hebei: implication for ~2500 Ma intra-continental rifting in the North China Craton. Precambrian Res. 208–211, 145–160.

Lu, L.Z., 1991. Metamorphic P–T–t path of the Archean granulite-facies terrains in Jining area, Inner Mongolia and its tectonic implications. Acta Petrol. Sin. 8, 1–12.

Lu, L.Z., Jin, S.Q., 1993. P–T–t paths and tectonic history of an early Precambrian granulite facies terrane, Jining district, southeastern Inner Mongolia, China. J. Metamorphic Geol. 11, 483–498.

Lu, L.Z., Xu, X.C., 1995. Metamorphism and deformation of metamorphic blocks in Yinshan–Northern Hebei Province. In: Lu, L.Z. (Ed.), Metamorphic Dynamics and Deformation of the Early Precambrian Rocks in the North China Craton. Changchun University of Science and Technology, Changchun, pp. 75–90. (in Chinese).

Lu, L.Z., Jin, S.Q., Xu, X.C., Liu, F.L., 1992. Petrogenesis and Mineralization of Khondalite Series in Southeastern Inner Mongolia. Jilin Science and Technology Press, Changchun.

Lu, L.Z., Xu, X.C., Liu, F.L., 1996. Early Precambrian Khondalite Series in North China. Changchun Publishing House, Changchun.

Lu, S.N., 1992. Chronology of Jixian Section of Middle-Upper Proterozoic strata. In: Li, Q.B. (Ed.), Symposium of Research on Modern Geology. Nanjing University Press, Nanjing, pp. 122–129.

Lu, S.N., Li, H.M., 1991. A precise U–Pb single zircon age determination for the volcanics of Dahongyu Formation, Changcheng System in Jixian. Bull. Chin. Acad. Geol. Sci. 22, 137–145 (in Chinese with English abstract).

Lu, S.N., Chen, Z.H., Xiang, Z.Q., 2008a. The Chronological Framework of Old Intrusions in the World Geological Park of Taishan. Geological Publishing House, Beijing, pp. 1–90 (in Chinese).

Lu, S.N., Zhao, G.C., Wang, H.C., Hao, G.J., 2008b. Precambrian metamorphic basement and sedimentary cover of the North China Craton: review. Precambrian Res. 160, 77–93.

Lu, X.P., Wu, F.Y., Guo, J.H., Wilde, S.A., Yang, J.H., Liu, X.M., et al., 2006. Zircon U–Pb geochronological constraints on the Paleoproterozoic crustal evolution of the Eastern Block in the North China Craton. Precambrian Res. 146, 138–164.

Luo, Y., Sun, M., Zhao, G.C., Li, S.Z., Xu, P., Ye, K., et al., 2004. LA-ICP-MS U–Pb zircon ages of the Liaohe Group in the Eastern Block of the North China Craton: constraints on the evolution of the Jiao-Liao-Ji Belt. Precambrian Res. 134, 349–371.

Luo, Y., Sun, M., Zhao, G.C., Li, S.Z., Xia, X.P., 2006. LA-ICP-MS U–Pb zircon geochronology of the Yushulazi Group in the Eastern Block, North China Craton. Int. Geol. Rev. 48, 828–840.

Luo, Y., Sun, M., Zhao, G.C., Ayers, J.C., Li, S.Z., Xia, X.P., et al., 2008. A comparison of U–Pb and Hf isotopic compositions of detrital zircons from the North and South Liaohe Group: constraints on the evolution of the Jiao-Liao-Ji Belt, North China Craton. Precambrian Res. 163, 279–306.

Ma, J., Wang, R.M., 1995. Discovery of coexisting kyanite + perthite assemblage in the Xuanhua-Chicheng high-pressure belt and its geological significance. Acta Petrol. Sin. 11, 273–278.

Ma, M.Z., Wan, Y.S., Santosh, M., Xu, Z.Y., Xie, H.Q., Dong, C.Y., et al., 2012. Decoding multiple tectonothermal events in zircons from single rock samples: SHRIMP zircon U–Pb data from the late Neoarchean rocks of Daqingshan, North China Craton. Gondwana Res. 22, 810–827.

Ma, M.Z., Xu, Z.Y., Zhang, L.C., Dong, C.Y., Dong, X.J., Liu, S.J., et al., 2013d. SHRIMP dating and Hf isotope analysis of zircons from the Early Precambrian basement in the Xiwulanbulang area, Wuchuan, Inner Mongolia. Acta Petrol. Sin. 29, 501–516.

Ma, X.D., Guo, J.H., Chen, L., Chu, Z.Y., 2010. Re–Os isotopic constraint to the age of komatiites in the Neoarchean Guyang greenstone belt, North China Craton. Chin. Sci. Bull. 55, 3197–3204.

Ma, X.D., Guo, J.H., Liu, F., Qian, Q., Fan, H.R., 2013a. Zircon U–Pb ages, trace elements and Nd–Hf isotopic geochemistry of Guyang sanukitoids and related rocks: implications for the Archean crustal evolution of the Yinshan Block, North China Craton. Precambrian Res. 230, 61–78.

Ma, X.D., Fan, H.R., Santosh, M., Guo, J.H., 2013b. Geochemistry and zircon U–Pb chronology of charnockites in the Yinshan Block, North China Craton: tectonic evolution involving Neoarchaean ridge subduction. Int. Geol. 55, 1688–1704.

Ma, X.D., Fan, H.R., Guo, J.H., 2013c. Neoarchean magmatism and metamorphism in the Yinshan Block: implications for the genesis of BIF and crustal evolution. Acta Petrol. Sin. 29, 2329–2339.

Ma, X.Y., Wu, Z.W., 1981. Early tectonic evolution of China. Precambrian Res. 14, 185–202.

Mao, D.B., Zhong, C.T., Chen, Z.H., Lin, Y.X., Li, H.M., Hu, X.D., 1999. Isotopic ages and geological implications of high-pressure mafic granulites in the northern Chengde area, Hebei Province, China. Acta Petrol. Sin. 15, 524–534.

Martin, H., Smithies, R.H., Rapp, R., Moyen, J.-F., Champion, D., 2005. An overview of adakite, tonalite–trondhjemite–granodiorite (TTG), and sanukitoid: relationships and some implications for crustal evolution. Lithos 79, 1–24.

Mason, R., 1990. Petrology of Metamorphic Rocks. Cambridge University Press, London, 230 pp.

Mei, H.L., 1994. P–T–t path and tectonic evolution of Paleoproterozoic metamorphic rocks in Zhongtiaoshan area. Geol. Rev. 40, 36–45 (in Chinese with English abstract).

Meng, E., Liu, F.L., Liu, J.H., Liu, P.H., Cui, Y., Liu, C.H., et al., 2013a. Zircon U–Pb and Lu–Hf isotopic constraints on Archean crustal evolution in the Liaonan Complex of northeast China. Lithos 177, 164–183.

Meng, E., Liu, F.L., Cui, Y., Cai, J., 2013b. Zircon U–Pb and Lu–Hf isotopic and whole-rock geochemical constraints on the protolith and tectonic history of the Changhai metamorphic supracrustal sequence in the Jiao-Liao-Ji Belt, southeast Liaoning Province, northeast China. Precambrian Res. 233, 297–315.

Miao, P.S., Zhang, Z.F., Zhang, J.Z., Zhao, Z.X., Xu, S.C., 1999. Discussion of Paleoproterozoic sedimentary sequences. Geol. China 18, 405–413 (in Chinese with English abstract).

Newton, R.C., Perkins, D., 1982. Thermodynamic calibration of geobarometers based the assemblage garnet–plagioclase–orthopyroxene (clinopyroxene)–quartz. Am. Mineral. 67, 203–222.

Nutman, A.P., Wan, Y.S., Du, L.L., Friend, C.R.L., Dong, C.Y., Xie, H.Q., et al., 2011. Multistage late Neoarchaean crustal evolution of the North China Craton, eastern Hebei. Precambrian Res. 189, 43–65.

Ouyang, J.P., Zhang, B.R., 1996. Geochemical evidence for the formation and evolution of the North Qinling microcontinent. Sci. China (D) 26 (Suppl), 42–48 (in Chinese).

O'Brien, P.J., Rötzler, J., 2003. High-pressure granulites: formation, recovery of peak conditions and implications for tectonics. J. Metamorphic Geol. 21, 3–20.

Pearce, J.A., Peate, D.W., 1995. Tectonic implications of the composition of volcanic arc magmas. Annu. Rev. Earth Planet. Sci. 23, 252–285.

Peng, P., 2005. Petrogenesis and Tectonic Significance of the ca. 1.8 Ga Mafic Dyke Swarms in the Central North China Craton (Ph.D. thesis). Chinese Academy of Sciences, Beijing, pp. 1–213.

Peng, P., Zhai, M.G., Zhang, H.F., Guo, J.H., 2005. Geochronological constraints on the Paleoproterozoic evolution of the North China craton: SHRIMP zircon ages of different types of mafic dikes. Int. Geol. Rev. 47, 492–508.

Peng, P., Zhai, M.G., Guo, J.H., Kusky, T., Zhao, T.P., 2007. Nature of mantle source contributions and crystal differentiation in the petrogenesis of the 1.78 Ga mafic dykes in the central North China craton. Gondwana Res. 12, 29–46.

Peng, P., Zhai, M.G., Ernst, R.E., Guo, J.H., Liu, F., Hu, B., 2008. A 1.78 Ga large igneous province in the North China craton: the Xiong'er Volcanic province and the North China dyke swarm. Lithos 101, 260–280.

Peng, P., Guo, J.H., Zhai, M.G., Bleeker, W., 2010. Paleoproterozoic gabbronoritic and granitic magmatism in the northern margin of the North China craton: evidence of crust–mantle interaction. Precambrian Res. 183, 635–659.

Peng, P., Guo, J.H., Windley, B.F., Li, X.H., 2011. Halaqin volcano-sedimentary succession in the central-northern margin of the North China Craton: products of Late Paleoproterozoic ridge subduction. Precambrian Res. 187, 165–180.

Peng, P., Guo, J.H., Windley, B.F., Liu, F., Chu, Z., Zhai, M.G., 2012a. Petrogenesis of Late Paleoproterozoic Liangcheng charnockites and S-type granites in the central-northern margin of the North China Craton: implications for ridge subduction. Precambrian Res. 222–223, 107–123.

Peng, Q.M., Palmer, M.R., 1995. The Paleoproterozoic boron deposits in eastern Liaoning, China—a metamorphosed evaporite. Precambrian Res. 72, 185–197.

Peng, T.P., Fan, W.M., Peng, P.X., 2012b. Geochronology and geochemistry of late Archean adakitic plutons from the Taishan granite–greenstone Terrain: implications for tectonic evolution of the eastern North China Craton. Precambrian Res. 208–211, 53–71.

Peng, T.P., Wilde, S.A., Fan, W.M., Peng, B.X., 2013a. Late Neoarchean potassic high Ba−Sr granites in the Taishan granite−greenstone terrane: petrogenesis and implications for continental crustal evolution. Chem. Geol. 344, 23−41.

Peng, T.P., Wilde, S.A., Fan, W.M., Peng, B.X., 2013b. Neoarchean siliceous high-Mg basalt (SHMB) from the Taishan granite−greenstone terrane, Eastern North China Craton: petrogenesis and tectonic implications. Precambrian Res. 228, 233−249.

Perkins, D., Chippera, S.J., 1985. Garnet−orthopyroxene−plagioclase−quartz barometry: refinement and application to English River subprovince and the Minnesota River Valley. Contrib. Mineral. Petrol. 89, 69−80.

Peucut, J.J., Jahn, B.M., Liu, D.Y., 1986. A precise zircon U−Pb age of tonalite from the Archean granite-greenstone belt in the Qingyuan area, Northeast China. International Conference on Precambrian Crustal Evolution, Beijing, Paper Collection 3, pp. 222−229.

Pidgeon, R.T., 1980. Isotopic ages of the zircons from the Archean granulite facies rocks, Eastern Hebei, China. Geol. Rev. 26, 198−207.

Plyusnina, L.P., 1982. Geothermometry and geobarometry of plagioclase−hornblende bearing assemblages. Contrib. Mineral. Petrol. 80, 140−146.

Polat, A., Kusky, T., Li, J.H., Fryer, B., Kerrich, R., Patrick, K., 2005. Geochemistry of Neoarchean (ca. 2.55−2.50) volcanic and ophiolitic rocks in the Wutaishan greenstone belt, central orogenic belt, North China craton: Implications for deodynamic setting and continental growth. Bull. Geol. Soc. Am. 117, 1387−1399.

Polat, A., Li, J.H., Fryer, B., Kusky, T., Kerrich, R., Gagnon, J., et al., 2006. Geochemical characteristics of the Neoarchean (2800−2700 Ma) Taishan greenstone belt, North China Craton: evidence for plume−craton interaction. Chem. Geol. 230, 60−87.

Puchtel, I.S., Hofmann, A.W., Mezger, K., Jochum, K.P., Shchipansky, A.A., Samsonov, A. V., 1998. Oceanic plateau model for continental crustal growth in the Archaean: a case study from the Kostomuksha greenstone belt, NW Baltic Shield. Earth Planet. Sci. Lett. 155, 57−74.

Qian, X.L., Li, J.H., 1999. The discovery of Neoarchean unconformity and its implication for continental cratonization of the north China Craton. Sci. China 42, 401−407 (Series D).

Qian, X.L., Cui, W.Y., Wang, S.Q., 1985. Evolution of the Inner Mongolia-eastern Hebei Archean granulite belt in the North China Craton. In: Department of Geology (Ed.), Records of Geological Research. Beijing University Press, Beijing, pp. 20−29.

Qiao, G.S., Wang, K.Y., Guo, Q.F., Zhang, G.C., 1987. Sm−Nd isotopic dating of the Paleoarchean rocks in Eastern Hebei, China. Sci. Geol. Sin. 22, 158−165 (in Chinese).

Qu, M., Guo, J.H., Lai, Y., Peng, P., Liu, F., 2012. Origin and geological significance of the 1.81 Ga hyalophane-rich pegmatite veins from the high-pressure granulite terrain in the Central Zone of North China Craton. Sci. China Earth Sci. 55, 193−203.

Rapp, R.P., Watson, E.B., Miller, C.F., 1991. Partial melting of amphibolite/eclogite and the origin of Archean trondhjemites and tonalites. Precambrian Res. 51, 1−25.

Ren, J.S., 1980. Tectonics and Evolution of China. Science Press, Beijing, pp. 64−75 (in Chinese).

Ren, F.G., Li, H.M., Yin, Y.J., Li, S.B., Ding, S.Y., Chen, Z.H., 2000. The upper chronological limit of the Xiong'er Group's volcanic rock series, and its geological significance. Progr. Precambrian Res. 23, 140−146 (in Chinese with English abstract).

Ren, Y.C., 2010. A Study on the Xihongshan Granite-Greenstone Terrane in Inner Mongolia. Jilin University, Changchun, pp. 1−69.

Rogers, J.J.W., Santosh, M., 2002. Configuration of Columbia, a Mesoproterozoic supercontinent. Gondwana Res. 5, 5−22.

Sajona, F.G., Maury, R.C., Bellon, H., Cotten, J., Defant, M., 1996. High field strength element enrichment of Pliocene–Pleistocene island arc basalts, Zamboanga Peninsula, Western Mindanao (Philippines). J. Petrol. 37, 693–726.

Santosh, M., 2010. Assembling North China Craton within the Columbia supercontinent: the role of double-sided subduction. Precambrian Res. 178, 149–167.

Santosh, M., Kusky, T.M., 2010. Origin of paired high pressure-ultrahigh-temperature orogens: a ridge subduction and slab window model. Terra Nova 22, 35–42.

Santosh, M., Sajeev, K., Li, J.H., 2006. Extreme crustal metamorphism during Columbia supercontinent assembly: evidence from North China Craton. Gondwana Res. 10, 256–266.

Santosh, M., Tsunogae, T., Li, J.H., 2007a. Discovery of sapphirine-bearing Mg–Al granulites in the North China Craton: implications for Paleoproterozoic ultrahigh temperature metamorphism. Gondwana Res. 11, 263–285.

Santosh, M., Wilde, S.A., Li, J.H., 2007b. Timing of Paleoproterozoic ultrahigh-temperature metamorphism in the North China Craton: evidence from SHRIMP U–Pb zircon geochronology. Precambrian Res. 159, 178–196.

Santosh, M., Tsunogae, T., Ohyama, H., Sato, K., Li, J.H., Liu, S.J., 2008. Carbonic metamorphism at ultrahigh-temperatures: evidence from North China Craton. Earth Planet. Sci. Lett. 266, 149–165.

Santosh, M., Sajeev, K., Li, J.H., Liu, S.J., Itaya, T., 2009a. Counterclockwise exhumation of a hot orogen: the Paleoproterozoic ultrahigh-temperature granulites in the North China Craton. Lithos 110, 140–152.

Santosh, M., Wan, Y., Liu, D., Chunyan, D., Li, J., 2009b. Anatomy of zircons from an ultrahot Orogen: the amalgamation of North China Craton within the supercontinent Columbia. J. Geol. 117, 429–443.

Santosh, M., Zhao, D.P., Kusky, T.M., 2010. Mantle dynamics of the Paleoproterozoic North China Craton: a perspective based on seismic tomography. J. Geodyn. 49, 39–53.

Santosh, M., Liu, S.J., Tsunogae, T., Li, J.H., 2012. Paleoproterozoic ultrahigh-temperature granulites in the North China Craton: implications for tectonic models on extreme crustal metamorphism. Precambrian Res. 222–223, 77–106.

Santosh, M., Liu, D.Y., Shi, Y.R., Liu, S.J., 2013. Paleoproterozoic accretionary orogenesis in the North China Craton: A SHRIMP zircon study. Precambrian. Res. 227, 29–54.

Sen, S.K., Bhattacharya, A., 1984. An orthopyroxene–garnet thermometry and its application to the Madras charnockites. Contrib. Mineral. Petrol. 88, 64–71.

Shen, B.F., Li, J.J., Mao, D.B., Li, S.B., Liu, Z.S., Zhang, W.J., et al., 1998. Geology and Metallogenic Genesis of the Jiapigou Gold Deposits in Jilin Province. Geological Publishing House, Beijing, pp. 1–175 (in Chinese with English abstract).

Shen, Q.H., Qian, X.L., 1995. Archean rock assemblages, episodes and tectonic evolution of China. Acta Geosci. Sin. 2, 113–120 (in Chinese with English abstract).

Shen, Q.H., Xu, H.F., Zhang, Z.Q., Gao, J.F., Wu, J.S., Ji, C.L., 1992. Precambrian Granulites in China. Geological Publishing House, Beijing, pp. 16–31 and 214–223.

Shen, Q.H., Song, B., Xu, H., Geng, Y.S., Shen, K., 2004. Emplacement and metamorphic ages of the Caiyu and Dashan igneous bodies, Yishui County, Shandong Province: zircon SHRIMP chronology. Geol. Rev. 50, 275–284.

Shen, Q.H., Zhao, Z.R., Song, B., Song, H.X., 2007. Geology, petrochemistry and SHRIMP zircon U-Pb dating of the Mashan and Xueshan granitoids in Yishui County, Shandong province. Geol. Rev. 53, 180–187.

Smithies, R.H., 2000. The Archaean tonalite–trondhjemite–granodiorite (TTG) series is not an analogue of Cenozoic adakite. Earth Planet. Sci. Lett. 182, 115–135.

Smithies, R.H., Champion, D.C., 2000. The Archean high-Mg diorite suite: links to tonalite-trondhjemite-granodirotie magmatism and implications for early Archaean crustal growth. Journal of Petrology 41, 1653–1671.

Smithies, R.H., Champion, D.C., Cassidy, K.F., 2003. Formation of Earth's early Archaean continental crust. Precambrian Res. 127, 89–101.

Song, B., Nutman, A.P., Liu, D.Y., Wu, J.S., 1996. 3800 to 2500 Ma crust in the Anshan area of Liaoning Province, northeastern China. Precambrian Res. 78, 79–94.

Song, H.X., Zhao, Z.R., Shen, Q.H., Song, B., 2009. Study on petrochemistry and hafnium isotope in Yishui Complex, Shandong Province. Acta Petrol. Sin. 25, 1872–1882 (in Chinese with English abstract).

Springer, W., Seck, H.A., 1997. Partial fusion of basic granulites at 5 to 15 kbar: implications for the origin of TTG magmas. Contrib. Mineral. Petrol. 127, 30–45.

Su, S.G., Gu, D.L., Zhu, G.X., 1999. Characteristics, age and petrogenesis of Yishui charnockites in Shandong. Earth Sci. 24, 57–62.

Su, W., Zhang, S., Huff, W.D., Li, H., Ettensohn, F.R., Chen, X., et al., 2008. SHRIMP U–Pb ages of K-bentonite beds in the Xiamaling Formation: implications for revised subdivision of the Meso- to Neoproterozoic history of the North China Craton. Gondwana Res. 14, 543–553.

Su, W., Li, H., Huff, W.D., Ettensohn, F.R., Zhang, S., Zhou, H., et al., 2010. SHRIMP U–Pb dating for a K-bentonite bed in the Tieling Formation, North China. Chin. Sci. Bull. 55, 3312–3323.

Sun, D.Y., Liu, Z.H., Zheng, C.Q., 1993a. Metamorphism and Tectonic Evolution of Early Precambrian Rocks in Fushun Area, the Northern Liaoning Province. Seismological Press, Beijing, pp. 90–120 (in Chinese).

Sun, D.Z., 1984. The Early Precambrian Geology of Eastern Hebei, China. Tianjin Science and Technology Press, Tianjin, pp. 24–34 (in Chinese with English abstract).

Sun, D.Z., Li, H.M., Lin, Y.X., Zhou, H.F., Zhao, F.Q., Tang, M., 1992c. Precambrian geochronology, chronotectonic framework and model of chronocrustal structure of the Zhongtiao Mountains. Acta Geol. Sin. 5, 23–37 (English edition).

Sun, D.Z., Hu, W.X., Tang, M., Zhao, F.Q., 1993b. The Geochronological Framework and Crustal Structures of Precambrian Basement in the Zhongtiaoshan Area. Geological Publishing House, Beijing, pp. 1–180 (in Chinese).

Sun, J.G., Lin, Q., Ge, W.C., 1992a. Deformation of Archean tonalitic intrusions in the Jinzhou area, Liaoning Province: implications for the tectonic setting of their emplacement. Jilin Geol. 4, 28–35.

Sun, M., Armstrong, R.L., Lambert, R.St.J., 1992b. Petrochemistry and Sr, Pb and Nd isotopic geochemistry of early Precambrian rocks, Wutaishan and Taihangshan areas, China. Precambrian Res. 56, 1–31.

Sun, S., Cong, B.L., Li, J.L., 1981. Meso-Neoproterozoic sedimentary basins in Henan and Shanxi Provinces. Sci. Geol. Sin. 16, 314–322.

Sun, S.S., McDonough, W.F., 1989. Chemical and isotopic systematics of oceanic basalts: implications for mantle composition and processes. In: Saunders, A.D., Norry, M.J. (Eds.), Magmatism in the Ocean Basins. Geological Society of London, pp. 313–345. , Special Publications 42.

Tam, P.Y., Zhao, G.C., Liu, F.L., Zhou, X.W., Sun, M., Li, S.Z., 2011. SHRIMP U–Pb zircon ages of high-pressure mafic and pelitic granulites and associated rocks in the Jiaobei massif: constraints on the metamorphic ages of the Paleoproterozoic Jiao-Liao-Ji Belt in the North China Craton. Gondwana Res. 19, 150–162.

Tam, P.Y., Zhao, G.C., Sun, M., Li, S.Z., Iizukac, Y., Ma, S.K., et al., 2012a. Metamorphic P–T path and tectonic implications of medium-pressure pelitic granulites from the Jiaobei massif in the Jiao-Liao-Ji Belt, North China Craton. Precambrian Res. 220–221, 177–191.

Tam, P.Y., Zhao, G.C., Zhou, X.W., Sun, M., Li, S.Z., Yin, C.Q., et al., 2012b. Metamorphic P−T path and implications of high-pressure pelitic granulites from the Jiaobei massif in the Jiao-Liao-Ji Belt, North China Craton. Gondwana Res. 22, 104−117.

Tam, P.Y., Zhao, G.C., Sun, M., Li, S.Z., Wu, M.L., Yin, C.Q., 2012c. Petrology and metamorphic P−T path of high-pressure mafic granulites from the Jiaobei massif in the Jiao-Liao-Ji Belt, North China Craton. Lithos 155, 94−109.

Tian, W., Liu, S.W., Liu, C.H., Yu, S.Q., Li, Q.G., Wang, Y.R., 2005. SHRIMP geochronology and geochemistry of the TTG rocks from the Shushui Complex in the Zhongtiaoshan area: geological implications. Prog. Nat. Sci. 15, 1476−1484.

Tian, Y., Zhao, D., Sun, R., Teng, J., 2009. Seismic imaging of the crust and upper mantle beneath the North China Craton. Phys. Earth Planet. Inter. 172, 169−182.

Tian, Y.Q., 1991. Geology and Mineralisation of the Wutai−Hengshan Greenstone Belt. Shanxi Science and Technology Press, Taiyuan, pp. 137−152 (in Chinese).

Tomlinson, K.Y., Stevenson, R.K., Hughes, D.J., Hall, R.P., Thurston, P.C., Henry, P., 1998. The Red Lake greenstone belt, Superior Province: evidence of plume-related magmatism at 3 Ga and evidence of an older enriched source. Precambrian Res. 89, 59−76.

Trap, P., Faure, M., Lin, W., Monié, P., 2007. Late Paleoproterozoic (1900−1800 Ma) nappe stacking and polyphase deformation in the Hengshan-Wutaishan area: implications for the understanding of the Trans-North-China Belt, North China Craton. Precambrian Res. 156, 85−106.

Trap, P., Faure, M., Lin, W., Bruguier, O., Monie, P., 2008. Contrasted tectonic styles for the Paleoproterozoic evolution of the North China Craton. Evidence for a ∼2.1 Ga thermal and tectonic event in the Fuping Massif. J. Struct. Geol. 30, 1109−1125.

Trap, P., Faure, M., Lin, W., Monie, P., Meffre, S., Melleton, J., 2009a. The Zanhuang Massif, the second and eastern suture zone of the Paleoproterozoic Trans-North China Orogen. Precambrian Res. 172, 80−98.

Trap, P., Faure, M., Lin, W., Meffre, S., 2009b. The Lüliang Massif: A Key Area for the Understanding of the Palaeoproterozoic Trans-North China Belt, North China Craton. Geol. Soc. Spec. Publ., London, Special Publication 338, pp. 99−125.

Trap, P., Faure, M., Lin, W., Augier, R., Fouassier, A., 2011. Syn-collisional channel flow and exhumation of Paleoproterozoic high pressure rocks in the Trans-North China Orogen: the critical role of partial-melting and orogenic bending. Gondwana Res. 20, 498−515.

Trap, P., Faure, M., Lin, W., Breton, N.L., Monie, P., 2012. The Paleoproterozoic evolution of the Trans-North China Orogen: toward a synthetic tectonic model. Precambrian Res. 222−223, 191−211.

Wan, Y.S., Zhang, Q., Song, T., 2003. SHRIMP ages of detrital zircons from the Changcheng System in the Ming Tombs area, Beijing: constraints on the protolith nature and maximum depositional age of the Mesoproterozoic cover of the North China Craton. Chin. Sci. Bull. 48, 2500−2506.

Wan, Y.S., Liu, D.Y., Song, B., Wu, J.S., Yang, C.H., Zhang, Z.Q., et al., 2005a. Geochemical and Nd isotopic compositions of 3.8 Ga meta-quartz dioritic and trondhjemitic rocks from the Anshan area and their geological significance. J. Asian Earth Sci. 4, 563−575.

Wan, Y.S., Song, B., Yang, C.H., Liu, D.Y., 2005b. Zircon SHRIMP U−Pb geochronology of Archean rocks from the Fushun-Qingyuan area, Liaoning Province and its geological significance. Acta Geol. Sin. 79, 78−87.

Wan, Y.S., Song, B., Liu, D.Y., Wilde, S.A., Wu, J.S., Shi, Y.R., et al., 2006a. SHRIMP U−Pb zircon geochronology of Palaeoproterozoic metasedimentary rocks in the North China Craton: evidence for a major late Palaeoproterozoic tectonothermal event. Precambrian Res. 149, 249−271.

Wan, Y.S., Wilde, S.A., Liu, D.Y., Yang, C.X., Song, B., Yin, X.Y., 2006b. Further evidence for ∼1.85 Ga metamorphism in the central Zone of the north China Craton: SHRIMP U−Pb dating of zircons from metamorphic rocks in the Lushan area, Henan Province. Gondwana Res. 9, 189−197.

Wan, Y.S., Liu, D.Y., Yin, X.Y., Wilde, S.A., Xie, L.W., Yang, Y.H., et al., 2007. SHRIMP geochronology and Hf isotope composition of zircons from the Tiejiashan granite and supracrustal rocks in the Anshan area, Liaoning Province. Acta Petrol. Sin. 23, 241–252.

Wan, Y.S., Liu, D.Y., Wang, S.J., Dong, C.Y., Yang, E.X., Wang, W., et al., 2010. Juvenile magmatism and crustal recycling at the end of the Neoarchean in Western Shandong Province, North China Craton: evidence from SHRIMP zircon dating. Am. J. Sci. 310, 1503–1552.

Wan, Y.S., Liu, D.Y., Wang, S.J., Yang, E.X., Wang, W., Dong, C.Y., et al., 2011a. ~2.7 Ga juvenile crust formation in the North China Craton (Taishan-Xintai area, western Shandong Province): further evidence of an understated event from U–Pb dating and Hf isotopic composition of zircon. Precambrian Res. 186, 169–180.

Wan, Y.S., Liu, D.Y., Dong, C.Y., Liu, S.J., Wang, Y.B., Yang, E.X., 2011b. U–Th–Pb behavior of zircons under high-grade metamorphic conditions: a case study of zircon dating of meta-diorite near Qixia, eastern Shandong. Geosci. Front. 2, 137–146.

Wan, Y.S., Wang, S.J., Liu, D.Y., Wang, W., Kröener, A., Dong, C.Y., et al., 2012a. Redefinition of depositional ages of Neoarchean supracrustal rocks in western Shandong Province, China: SHRIMP U–Pb zircon dating. Gondwana Res. 21, 768–784.

Wan, Y.S., Dong, C.Y., Liu, D.Y., Kröner, A., Yang, C.H., Wang, W., et al., 2012b. Zircon ages and geochemistry of late Neoarchean syenogranites in the North China Craton: a review. Precambrian Res. 222–223, 265–289.

Wan, Y.S., Liu, D.Y., Nutman, A., Zhou, H.Y., Dong, C.Y., Yin, X.Y., et al., 2012c. Multiple 3.8–3.1 Ga tectono-magmatic events in a newly discovered area of ancient rocks (the Shengousi Complex), Anshan, North China Craton. J. Asian Earth Sci. 54–55, 18–30.

Wan, Y.S., Zhang, Y.H., Williams, I.S., Dong, C.Y., Fan, Y.L., Shi, Y.R., et al., 2013. Extreme zircon O isotopic compositions from 3.8 to 2.5 Ga magmatic rocks from the Anshan area, North China Craton. Chem. Geol. 352, 108–124.

Wang, A.D., Liu, Y.C., Gu, X.F., Hou, Z.H., Song, B., 2012b. Late-Neoarchean magmatism and metamorphism at the southeastern margin of the North China Craton and their tectonic implications. Precambrian Res. 220–221, 65–79.

Wang, F., Li, X.P., Chu, H., Zhao, G.C., 2011a. Petrology and metamorphism of khondalites from Jining Complex in the North China Craton. Int. Geol. Rev. 53, 212–229.

Wang, J., Wu, Y.B., Gao, S., Peng, M., Liu, X.C., Zhao, L.S., et al., 2010a. Zircon U–Pb and trace element data from rocks of the Huai'an Complex: new insights into the late Paleoproterozoic collision between the Eastern and Western blocks of the North China Craton. Precambrian Res. 178, 59–71.

Wang, K.Y., Li, J.L., Liu, L.Q., 1991a. Petrogenesis of the Fuping grey gneisses. Sin. Geol. Sci. 11, 254–267 (in Chinese).

Wang, K.Y., Li, J.L., Hao, J., Li, J.H., Zhou, S.P., 1996. The Wutaishan mountain belt within the Shanxi Province, Northern China: a record of late Archean collision tectonics. Precambrian Res. 78, 95–103.

Wang, K.Y., Li, J.L., Hao, J., Li, J.H., Zhou, S.P., 1997. Late Archean mafic-ultramafic rocks from the Wuataishan, Shanxi Province: a possible ophiolite melange. Acta Petrol. Sin. 13, 139–151.

Wang, K.Y., Wang, Z., Yu, L., Fan, H., Wilde, S.A., Cawood, P.A., 2001. Evolution of Archaean greenstone belt in the Wutaishan region, North China: constraints from SHRIMP zircon U-Pb and other geochronological and isotope information. In: Cassidy, K.F. (Ed.), Proceedings of the Fourth International Archaean Symposium 2001, Extended Abstracts. AGSO, Geoscience Australia, Record 37, 104–105.

Wang, R.M., Chen, J.J., Chen, F., 1991b. Significance of grey gneiss and high-pressure granulite from Hengshan. Acta Petrol. Sin. 4, 119–131.

Wang, R.M., Lai, X.Y., Dong, W.D., Ma, J., Tang, B., 1994. Evidence for a Late Archaean collision belt in Western Hebei. In: Qian, X.L., Wang, R.M. (Eds.), Geological Evolution of the Granulite Terrain in the North Part of the North China Craton. Seismological Press, Beijing, pp. 7–20. (in Chinese).

Wang, R.M., Ni, Z.Y., Yuan, J.M., Tong, Y., 2002. More on the evidence of the paleostylolitic zone in northern Hebei Province. Acta Petrol. Mineral. 21, 327–335.

Wang, R.M., Wan, Y.S., Cheng, S.H., Feng, Y.G., 2009a. Modern-style subduction processes in the Archean: evidence from the Shangyi Complex in North China Craton. Acta Geol. Sin. 83, 525–543 (English edition).

Wang, S.J., Wan, Y.S., Zhang, C.J., Yang, A.X., Song, Z.Y., Wang, L.F., et al., 2008. New progress made in early Precambrian geology of the Luxi area. Land and Resources of Shandong Province 24, 10–20 (in Chinese).

Wang, W., Wang, S.J., Liu, D.Y., Li, P.Y., Dong, C.Y., Xie, H.Q., et al., 2010b. Formation age of the Neoarchean Jining Group (banded iron formation) in western Shandong Province: constraints from SHRIMP U–Pb zircon dating. Acta Petrol. Sin. 26, 1175–1181.

Wang, W., Liu, S.W., Xiang, B., Yang, P.T., Li, Q.G., Zhang, L.F., 2011b. Geochemistry and zircon U–Pb–Hf isotopic systematics of the Neoarchean Yixian–Fuxin greenstone belt, northern margin of the North China Craton: implications for petrogenesis and tectonic setting. Gondwana Res. 20, 64–81.

Wang, W., Liu, S.W., Wilde, S.A., Li, Q.G., Zhang, J., Xiang, B., et al., 2012a. Petrogenesis and geochronology of Precambrian granitoid gneisses in Western Liaoning Province: constraints on Neoarchean to early Paleoproterozoic crustal evolution of the North China Craton. Precambrian Res. 222–223, 290–311.

Wang, W., Yang, E., Zhai, M., Wang, S., Santosh, M., Du, L., et al., 2013a. Geochemistry of ~2.7 Ga basalts from Taishan area: constraints on the evolution of early Neoarchean granite-greenstone belt in western Shandong Province, China. Precambrian Res. 224, 94–109.

Wang, W., Liu, S.W., Santosh, M., Bai, X., Li, Q.L., Yang, P.T., et al., 2013b. Zircon U–Pb–Hf isotopes and whole-rock geochemistry of granitoid gneisses in the Jianping gneissic terrane, Western Liaoning Province: constraints on the Neoarchean crustal evolution of the North China Craton. Precambrian Res. 224, 184–221.

Wang, Y.J., Fan, W.M., Zhang, Y., Guo, F., 2003. Structural evolution and $^{40}Ar/^{39}Ar$ dating of the Zanhuang metamorphic domain in the North China Craton: constraints on Paleoproterozoic tectonothermal overprinting. Precambrian Res. 122, 159–182.

Wang, Y.J., Fan, W.M., Zhang, Y.H., Guo, F., Zhang, H., Peng, P., 2004b. Geochemical $^{40}Ar/^{39}Ar$ geochronological and Sr–Nd isotopic constraints on the origin of Paleoproterozoic mafic dikes from the southern Taihang Mountains and implications for the ca. 1800 Ma event of the North China Craton. Precambrian Res. 135, 55–77.

Wang, Y.J., Zhang, Y.Z., Zhao, G.C., Fan, W.M., Xia, X.P., Zhang, F.F., et al., 2009b. Zircon U–Pb geochronological and geochemical constraints on the petrogenesis of the Taishan sanukitoids (Shandong): implications for Neoarchean subduction in the Eastern Block, North China Craton. Precambrian Res. 174, 273–286.

Wang, Z.H., 2010. Reply to the comment by Zhao et al. on: Tectonic evolution of the Hengshan–Wutai–Fuping complexes and its implication for the Trans-North China Orogen [Precambrian Res. 170 (2009) 73–87]. Precambrian Res. 176, 99–104.

Wang, Z.H., Wilde, S.A., Wang, K.Y., Yu, L.J., 2004a. A MORB-arc basalt-adakite association in the 2.5 Ga Wutai greenstone belt: late Archean magmatism and crustal growth in the North China Craton. Precambrian Res. 131, 323–343.

Wang, Z.H., Wilde, S.A., Wan, J.L., 2010c. Tectonic setting and significance of 2.3-2.1 Ga magmatic events in the Trans-North China Orogen: new constraints from the Yanmenguan mafic-ultramafic intrusion in the Hengshan-Wutai-Fuping area. Precambrian Res. 178, 27–42.

Wells, P.R.A., 1980. Thermal models for magmatic accretion and subsequent metamorphism of continental crust. Earth Planet. Sci. Lett. 46, 253–265.

Wilde, S., 2002. SHRIMP U–Pb zircon ages of the Wutai Complex. In: Kröner, A., Zhao, G.C., Wilde, S.A., Zhai, M.G., Passchier, C.W., Sun, M., et al.,A Late Archaean to Palaeoproterozoic Lower to Upper Crustal Section in the Hengshan-Wutaishan Area of North China. Guidebook for Penrose Conference Field Trip, September 2002. Chinese Academy of Sciences, Beijing, pp. 32–34.

Wilde, S.A., Zhao, G.C., 2005. Archean to Paleoproterozoic evolution of the North China Craton. J. Asian Earth Sci. 24, 519–522.

Wilde, S.A., Cawood, P., Wang, K.Y., 1997. The relationship and timing of granitoid evolution with respect to felsic volcanism in the Wutai Complex, North China Craton. In: Qian, X.L., You, Z.D., Halls, H.C. (Eds.), Proceedings of the 30th IGC. Precambrian Geology-Metamorphic. Petrology 17, 75–88.

Wilde, S.A., Cawood, P.A., Wang, K.Y., 1998. SHRIMP U–Pb data of granites and gneisses in the Taihangshan–Wutaishan area: implications for the timing of crustal growth in the North China Craton. Chin. Sci. Bull. 43, 144.

Wilde, S.A., Zhao, G.C., Sun, M., 2002. Development of the North China Craton during the Late Archaean and its final amalgamation at 1.8 Ga; some speculations on its position within a global Palaeoproterozoic Supercontinent. Gondwana Res. 5, 85–94.

Wilde, S.A., Cawood, P.A., Wang, K.Y., Nemchin, A., Zhao, G.C., 2004a. Determining Precambrian crustal evolution in China: a case-study from Wutaishan, Shanxi Province, demonstrating the application of precise SHRIMP U–Pb geochronology. In: Malpas, J., Fletcher, C.J., Aitchison, J.C., Ali, J. (Eds.), Aspects of the Tectonic Evolution of China, 226. Geological Society Special Publication, London, pp. 5–26.

Wilde, S.A., Zhao, G.C., Wang, K.Y., Sun, M., 2004b. First precise SHRIMP U–Pb zircon ages for the Hutuo Group, Wutaishan: further evidence for the Palaeoproterozoic amalgamation of the North China Craton. Chin. Sci. Bull. 49, 83–90.

Wilde, S.A., Cawood, P.A., Wang, K.Y., Nemchin, A., 2005. Granitoid evolution in the late Archean Wutai Complex, North China Craton. J. Asian Earth Sci. 24, 597–613.

Wilde, S.A., Valley, J.W., Kita, N.T., Cavosie, A.J., Liu, D.-Y., 2008. SHRIMP U–Pb and CAMECA 1280 oxygen isotope results from ancient detrital zircons in the Caozhuang quartzite, Eastern Hebei, North China Craton: evidence for crustal reworking 3.8 Ga ago. Am. J. Sci. 308, 185–199.

Wu, C.H., Zhong, C.T., 1998. The Paleoproterozoic SW–NE collision model for the central North China Craton. Prog. Precambrian Res. 21, 28–50 (in Chinese).

Wu, C.H., Li, S.X., Gao, J.F., Xu, X.C., 1986. The Archean to Palaeoproterozoic basement rocks in the North China. In: Dong, S.B. (Ed.), Relationships Between Metamorphism and Crustal Evolution in China. Geological Publishing House, Beijing, pp. 53–99. (in Chinese with English abstract).

Wu, C.H., Sun, M., Li, H.M., Zhao, G.C., Xia, X.P., 2006. LA-ICP-MS U–Pb zircon ages of the khondalites from the Wulashan and Jining high-grade terrain in northern margin of the North China Craton: constraints on sedimentary age of the khondalite. Acta Petrol. Sin. 22, 2639–2654.

Wu, F.Y., Ge, W.C., Sun, D.Y., Lin, Q., Zhou, Y., 1997. The Sm–Nd, Rb–Sr isotopic ages of the Archean granites in southern Jilin Province. Acta Petrol. Sin. 13, 499–506 (in Chinese with English abstract).

Wu, F.Y., Yang, J.H., Liu, X.M., Li, T.S., Xie, L.W., Yang, Y.H., 2005a. Hf isotopes of the 3.8 Ga zircons in eastern Hebei Province, China: implications for early crustal evolution of the NCC. Chin. Sci. Bull. 50, 2473–2480.

Wu, F.Y., Zhao, G.C., Wilde, S.A., Sun, D.Y., 2005b. Nd isotopic constraints on crustal formation in the North China Craton. J. Asian Earth Sci. 24, 523–545.

Wu, F.Y., Li, X.H., Zheng, Y.F., Gao, S., 2007. Lu–Hf isotopic systematics and their applications to petrology. Acta Petrol. Sin. 23, 185–220.

Wu, F.Y., Zhang, Y.B., Yang, J.H., Xie, L.W., Yang, Y.H., 2008. Zircon U–Pb and Hf isotopic constraints on the Early Archean crustal evolution in Anshan of the North China Craton. Precambrian Res. 167, 339–362.

Wu, F.Y., Zhang, Y.B., Yang, J.H., Xie, L.W., Yang, Y.H., 2009. Are there any 3.8 Ga rock at Anshan in the North China Craton? Reply to comments on Zircon U–Pb and Hf isotopic constraints on the Early Archean crustal evolution in Anshan of the North China Craton by Nutman et al. Reply. Precambrian Res. 172, 361–363.

Wu, J.S., Geng, Y.S., Shen, Q.H., Liu, D.Y., Li, Z.L., Zhao, D.M., 1991. The Early Precambrian Significant Geological Events in the North China Craton. Geological Publishing House, Beijing, pp. 1–115 (in Chinese with English abstract).

Wu, J.S., Geng, Y.S., Shen, Q.H., 1998. Archaean Geology Characteristics and Tectonic Evolution of Sino-Korea Paleo-Continent. Geological Publishing House, pp. 192–211 (in Chinese).

Wu, K.K., Zhao, G.C., Sun, M., Yin, C.Q., He, Y.H., Tam, P.Y., 2013d. Metamorphism of the Northern Liaoning Complex: implications for the tectonic evolution of the Late Archean basement of the Eastern Block, North China Craton. Geosci. Front. 4, 305–320.

Wu, M.L., Zhao, G.C., Sun, M., Yin, C.Q., Li, S.Z., Tam, P.Y., 2012. Petrology and P–T path of the Yishui mafic granulites: implications for tectonothermal evolution of the Western Shandong Complex in the Eastern Block of the North China Craton. Precambrian Res. 222–223, 312–324.

Wu, M.L., Zhao, G.C., Sun, M., Li, S.Z., Bao, Z.A., Tam, P.Y., et al., 2013a. Zircon U–Pb geochronology and Hf isotopes of major lithologies from the Jiaodong Terrane: implications for the crustal evolution of the Eastern Block of the North China Craton. Lithos (in press).

Wu, M.L., Zhao, G.C., Sun, M., Li, S.Z., He, Y.H., Bao, Z.A., 2013b. Zircon U-Pb geochronology and Hf isotopes of major lithologies from the Yishui terrane: implications for the crustal evolution of the Eastern Block, North China Craton. Lithos 170–171, 164–178.

Wu, M.L., Zhao, G.C., Sun, M., Bao, Z.A., Tam, P.Y., He, Y.H., 2013c. Tectonic affinity and reworking of the Archean Jiaodong Terrane in the Eastern Block of the North China Craton: evidence from LA-ICP-MS U–Pb zircon ages. Geol. Mag.. Available from: http://dx.doi.org/10.1017/S0016756813000721.

Xia, X.P., Sun, M., Zhao, G.C., Luo, Y., 2006a. LA-ICP-MS U–Pb geochronology of detrital zircons from the jining complex, North China Craton and its tectonic significance. Precambrian Res. 144, 199–212.

Xia, X.P., Sun, M., Zhao, G.C., Wu, F.Y., Xu, P., Zhang, J.H., et al., 2006b. U–Pb and Hf isotopic study of detrital zircons from the Wulashan khondalites: constraints on the evolution of the Ordos Terrane, Western Block of the North China Craton. Earth Planet. Sci. Lett. 241, 581–593.

Xia, X.P., Sun, M., Zhao, G.C., Wu, F.Y., Xu, P., Zhang, J., et al., 2006c. U–Pb and Hf isotope study of detrital zircons from the Wanzi supracrustals: constraints on the tectonic setting and evolution of the Fuping Complex, Trans-North China Orogen. Acta Geol. Sin. 80, 844–863 (English edition).

Xia, X.P., Sun, M., Zhao, G.C., Wu, F.Y., Xu, P., Zhang, J.S., 2008. Paleoproterozoic crustal growth events in the Western Block of the North China Craton: evidence from detrital zircon Hf and whole rock Sr–Nd isotopes of the khondalites in the Iining Complex. Am. J. Sci. 308, 304–327.

Xia, X.P., Sun, M., Zhao, G.C., Wu, F.Y., Xu, P., Zhang, J.S., 2009. Detrital zircon U–Pb age and Hf isotope study of the khondalite in Trans-North China Orogen and its tectonic significance. Geol. Mag. 146, 701–716.

Xiao, L.L., Jiang, Z.S., Wang, G.D., Wan, Y.S., Wang, T., Wu, C.M., 2011a. Metamorphic reaction textures and metamorphic P–T–t loops of the Precambrian Zanhuang metamorphic complex, Hebei, North China. Acta Petrol. Sin. 27, 980–1002.

Xiao, L.L., Wu, C.M., Zhao, G.C., Guo, J.H., Ren, L.D., 2011b. Metamorphic P–T paths of the Zanhuang amphibolites and metapelites: constraints on the tectonic evolution of the Paleoproterozoic Trans-North China Orogen. Int. J. Earth Sci. 100, 717–739.

Xu, H.F., Cheng, Y.Q., Zhuang, Y.X., 1992. Granite–Greenstone Belt of the Western Shandong Province. Geological Publishing House, Beijing (in Chinese).

Xu, Z.Y., Liu, Z.H., Yang, Z.S., 2001. The composition and characteristics of the early Precambrian metamorphic strata in the Daqingshan region, Inner Mongolia. World Geol. 20, 209–218.

Xu, Z.Y., Liu, Z.H., Yang, Z.S., 2002. The structural architectures of the khondalites in the Daqingshan area, Inner Mongolia. J. Jilin Univ. 32, 313–318 (Earth Science edition).

Yang, C.H., Du, L.L., Ren, L.D., Song, H.X., Xie, H.Q., Liu, Z.X., 2011. The age and petrogenesis of the Xuting granite in the Zanhuang Complex, Hebei Province: constraints on the structural evolution of the Trans-North China Orogen, North China Craton. Acta Petrol. Sin. 27, 1003–1016.

Yang, J.H., Wu, F.Y., Liu, X.M., Xie, L.W., 2005. Zircon U–Pb ages and Hf isotopes and their geological significance of the Miyun rapakivi granites from Beijing, China. Acta Petrol. Sin. 21, 1633–1644 (in Chinese with English abstract).

Yang, J.H., Wu, F.Y., Wilde, S.A., Zhao, G.C., 2008. Petrogenesis and geodynamics of Late Archean magmatism in the eastern North China Craton: geochronological, geochemical and Nd–Hf isotopic evidence. Precambrian Res. 167, 125–149.

Yang, Y., 1990. Characteristics, tectonic setting of volcanics in the Xiong'er Group along the southern margin of the North China Craton. Acta Petrol. Sin. 6, 20–29 (in Chinese with English abstract).

Yang, Z.S., Xu, Z.Y., Liu, Z.H., 2000. The Khondalite event and Archean crust structural evolvement. Prog. Precambrian Res. 23, 206–212.

Yin, C.Q., 2010. Metamorphism of the Helanshan-Qianlishan Complex and its Implications for Tectonic Evolution of the Khondalite Belt in the Western Block, North China Craton (Ph.D. thesis). The University of Hong Kong, 233 p.

Yin, C.Q., Zhao, G.C., Sun, M., Xia, X.P., Wei, C.J., Leung, W.H., 2009. LA-ICP-MS U–Pb zircon ages of the Qianlishan Complex: constraints on the evolution of the Khondalite Belt in the Western block of the North China Craton. Precambrian Res. 174, 78–94.

Yin, C.Q., Zhao, G.C., Guo, J.H., Sun, M., Xia, X.P., Zhou, X.W., et al., 2011. U–Pb and Hf isotopic study of zircons of the Helanshan Complex: constraints on the evolution of the Khondalite Belt in the Western Block of the North China Craton. Lithos 122, 25–38.

Yin, C.Q., Zhao, G.C., Wei, C.J., Sun, M., Guo, J.H., Zhou, X.W., 2014a. High-pressure Pelitic Granulites from the Helanshan Complex in the Khondalite Belt, North China Craton: metamorphic P–T path and tectonic implications. Am. J. Sci. (in press).

Yin, C.Q., Zhao, G.C., Wei, C.J., Sun, M., Guo, J.H., 2014b. Metamorphism and partial melting of high-pressure pelitic granulites from the Qianlishan Complex: constraints on the tectonic evolution of the Khondalite Belt in the North China Craton. Precambrian Res. (in press).

Yu, J.H., Fu, H.Q., Zhang, F.L., Wan, F.Z., Haapala, I., Ramo, O.T., et al., 1996. Anorogenic Rapakivi Granites and Related Rocks in the Northern of North China Craton. Science and Technology Press, Beijing.

Yu, J.H., Wang, D.Z., Wang, X.Y., 1997. Ages of the Lüliang Group and its main metamorphism in the Lüliang Mountains, Shanxi: evidence from single-grain zircon U–Pb ages. Geol. Rev. 43, 403–408.

Zhai, M.G., 2004. Precambrian geological events in the North China Craton. In: Malpas, J., Fletcher, C.J.N., Ali, J.R., Aitchison, J.C. (Eds.), Tectonic Evolution of China. Geological Society of London, pp. 57–72. , Special Publication 226.

Zhai, M.G., 2009. Two kinds of granulites (HT-HP and HT-UHT) in the North China Craton: their genetic relations and geodynamic implications. Acta Petrol. Sin. 25, 1753–1771.

Zhai, M.G., 2011. Cratonization and the ancient North China Continent: a summary and review. Sci. China 54, 1110–1120 (Series D Earth Science).

Zhai, M.G., Liu, W.J., 2001. An oblique cross-section of Precambrian crust in the North China Craton. Phys. Chem. Earth (A) 26, 781–792.

Zhai, M.G., Liu, W.J., 2003. Paleoproterozoic tectonic history of the North China Craton: a review. Precambrian Res. 122, 183–199.

Zhai, M.G., Peng, P., 2007. Paleoproterozoic events in North China Craton. Acta Petrol. Sin. 23, 2665–2682 (in Chinese with English abstract).

Zhai, M.G., Santosh, M., 2011. The early Precambrian odyssey of North China Craton: a synoptic overview. Gondwana Res. 20, 6–25.

Zhai, M.G., Yang, R.Y., Lu, W.J., Zhou, J., 1985. Geochemistry and evolution of the Qingyuan Archaean granite–greenstone, terrain, NE, China. Precambrian Res. 27, 37–62.

Zhai, M.G., Windley, B.F., Sills, J.D., 1990. Archaean gneisses, amphibolites, banded iron-formation from Anshan area of Liaoning, NE China: their geochemistry, metamorphism and petrogenesis. Precambrian Res. 46, 195–216.

Zhai, M.G., Guo, J.H., Yan, Y.H., 1992. Discovery and preliminary study of the Archean high-pressure granulites in the North China. Sci. China 12B, 1325–1330.

Zhai, M.G., Guo, J.H., Li, H.H., Yan, Y.H., Li, Y.G., 1995. Discovery of retrograded eclogites in the Archaean North China Craton. Chin. Sci. Bull. 40, 1590–1594.

Zhai, M.G., Bian, A.G., Zhao, T.P., 2000. The amalgamation of the supercontinent of North China Craton at the end of Neo-Archaean and its breakup during late Palaeoproterozoic and Mesoproterozoic. Sci. China 43, 219–232 (Series D Earth Science).

Zhai, M.G., Shao, J.A., Hao, J., Peng, P., 2003. Geological signature and possible position of the North China block in the Supercontinent Rodinia. Gondwana Res. 6, 171–183.

Zhai, M.G., Guo, J.H., Liu, W.J., 2005. Neoarchean to Paleoproterozoic continental evolution and tectonic history of the North China Craton. J. Asian Earth Sci. 24, 547–561.

Zhai, M.G., Li, T.S., Peng, P., Hu, B., Liu, F., Zhang, Y.B., et al., 2010. Precambrian key tectonic events and evolution of the North China Craton. In: Kusky, T.M., Zhai, M.G., Xiao, W.J. (Eds.), The Evolving Continents. Geological Society of London, pp. 235–262. Special Publication 338.

Zhang, B.R., Zhang, H.F., Zhao, Z.D., Lin, W.L., 1996b. Geochemical subdivision and evolution of the lithosphere in East Qinling and adjacent regions: implications for tectonics. Sci. China (D) 39, 245–255.

Zhang, F.Q., Liu, J.Z., Ouyang, Z.Y., 1998. Tectonic framework of greenstones in the basement of the North China Craton. Acta Geophys. Sin. 41, 99–107.

Zhang, G.W., 1989. Formation and Evolution of the Qinling Orogen. Northwest University Press, Xi'an, 199 pp. (in Chinese).

Zhang, G.W., Zhang, Z.Q., Dong, Y.P., 1995. Nature of main tectono-lithostratigraphic units of the Qinling Orogen: implications for the tectonic evolution. Acta Petrol. Sin. 11, 101–114.

Zhang, G.W., Meng, Q.R., Yu, Z.P., 1996a. Orogenic processes and geodynamics of the Qinling orogen. Sci. China (D) 26, 193–200.

Zhang, H.F., Ying, J.F., Santosh, M., Zhao, G.C., 2012a. Episodic growth of Precambrian lower crust beneath the North China Craton: a synthesis. Precambrian Res. 222–223, 255–264.

Zhang, J., Zhao, G.C., Sun, M., Wilde, S.A., Li, S.Z., Liu, S.W., 2006a. High-pressure mafic granulites in the Trans-North China Orogen: tectonic significance and age. Gondwana Res. 9, 349–362.

Zhang, J., Zhao, G.C., Li, S.Z., Sun, M., Liu, S.W., Wilde, S.A., et al., 2007a. Deformation history of the Hengshan Complex: implications for the tectonic evolution of the Trans-North China Orogen. J. Struct. Geol. 29, 933–949.

Zhang, J., Zhao, G.C., Li, S.Z., Sun, M., Liu, S.W., Yin, C.Q., 2009a. Deformational history of the Fuping Complex and new U–Th–Pb geochronological constraints: implications for the tectonic evolution of the Trans-North China Orogen. J. Struct. Geol. 31, 177–193.

Zhang, J., Li, J.Y., Liu, J.F., Li, Y.F., Qu, J.F., Feng, Q.W., 2012b. The relationship between the Alxa Block and the North China Plate during the Early Paleozoic: new information from the Middle Ordovician detrital zircon ages in the eastern Alxa Block. Acta Petrol. Sin. 28, 2912–2934.

Zhang, J., Zhao, G.C., Li, S.Z., Sun, M., Liu, S.W., 2012c. Structural and aeromagnetic studies of the Wutai Complex: implications for the tectonic evolution of the Trans-North China Orogen. Precambrian Res. 222–223, 212–229.

Zhang, J.S., Dirks, H.G.M., Passchier, C.W., 1994. Extensional collapse and uplift in a polymetamorphic granulite terrain in the Archean and Paleoproterozoic of North China. Precambrian Res. 67, 37–57.

Zhang, L.C., Zhai, M.G., Zhang, X.J., Xiang, P., Dai, Y.P., Wang, C.L., et al., 2012d. Formation age and tectonic setting of the Shirengou Neoarchean banded iron deposit in eastern Hebei Province: constraints from geochemistry and SIMS zircon U–Pb dating. Precambrian Res. 222–223, 325–338.

Zhang, Q.S., Yang, Z.S., 1988. Early Crust and Mineral Deposits of Liaodong Peninsula, China. Geological Publishing House, Beijing.

Zhang, S.H., Liu, S.W., Zhao, Y., Yang, J.H., Song, B., Liu, X.M., 2007b. The 1.75–1.68 Ga anorthosite-mangerite-alkali granitoid-rapakivi granite suite from the northern North China Craton: magmatism related to a Paleoproterozoic orogen. Precambrian Res. 155, 287–312.

Zhang, S.H., Zhao, Y., Yang, Z.Y., He, Z.F., Wu, H., 2009b. The 1.35 Ga diabase sills from the northern North China Craton: implications for breakup of the Columbia (Nuna) supercontinent. Earth Planet. Sci. Lett. 288, 588–600.

Zhang, S.H., Zhao, Y., Santosh, M., 2012e. Mid-Mesoproterozoic bimodal magmatic rocks in the northern North China Craton: implications for magmatism related to breakup of the Columbia supercontinent. Precambrian Res. 222–223, 339–367.

Zhang, S.H., Li, Z.X., Evans, D.A.D., Wu, H.C., Li, H.Y., Dong, J., 2012f. Pre-Rodinia supercontinent Nuna shaping up: a global synthesis with new Paleomagnetic results from North China. Earth Planet. Sci. Lett. 353–354, 145–155.

Zhang, S.H., Zhao, Y., Ye, H., Hu, J.M., Wu, F., 2013. New constraints on ages of the Chuanlinggou and Tuanshanzi formations of the Changcheng System in the Yan-Liao area in the northern North China Craton. Acta Petrol. Sin. 29, 2481–2490.

Zhang, X.J., Zhang, L.C., Xiang, P., Wan, B.O., Pirajno, F., 2011. Zircon U–Pb age, Hf isotopes and geochemistry of Shuichang Algoma-type banded iron-formation, North China Craton: constraints on the ore-forming age and tectonic setting. Gondwana Res. 20, 137–148.

Zhang, Y.Q., 2004. Zircon U–Pb age of the quartz diorite in North Daqing mountains, central Inner Mongolia. Geol. Miner. Resour. South China 4, 22–27.

Zhang, Y.Q., Wang, T., Jia, H.Y., Zhang, Z.X., 2003. U–Pb ages of zircons from the Xi Ulanbulang hypersthene-plagioclase granulite in the north Daqing mountains, central Inner Mongolia. Geol. China 30, 394–399.

Zhang, Y.Q., Zhang, Y.K., Zheng, B.J., Xu, G.Q., Han, J.G., Mu, L.J., et al., 2006b. Geological character and significance of adakite and TTG in the Xiaonangou-Mingxinggou area in the central Inner Mongolia. Acta Petrol. Sin. 22, 2762−2768.

Zhang, Y.X., Yan, H.Q., Wang, K.D., Li, F.Y., 1980. Komatiites from the Qianxi Group in the eastern Hebei Province, China. J. Changchun Univ. Sci. Tech. 10, 1−8 (in Chinese).

Zhao, G.C., 2001. Palaeoproterozoic assembly of the North China Craton. Geol. Mag. 138, 87−91.

Zhao, G.C., 2009. Metamorphic evolution of major tectonic units in the basement of the North China Craton: key issues and discussion. Acta Petrol. Sin. 25, 1772−1792.

Zhao, G.C., Cawood, P.A., 2012. Precambrian geology of China. Precambrian Res. 222−223, 13−54.

Zhao, G.C., Guo, J.H., 2012. Precambrian geology of China: a preface. Precambrian Res. 222−223, 1−12.

Zhao, G.C., Zhai, M.G., 2013. Lithotectonic elements of Precambrian basement in the North China Craton: review and tectonic implications. Gondwana Res. 23, 1207−1240.

Zhao, G.C., Wilde, S.A., Cawood, P.A., 1998. Thermal evolution of Archean basement rocks from the eastern part of the North China Craton and its bearing on tectonic setting. Int. Geol. Rev. 40, 706−721.

Zhao, G.C., Wilde, S.A., Cawood, P.A., Lu, L.Z., 1999a. Tectonothermal history of the basement rocks in the western zone of the North China Craton and its tectonic implications. Tectonophysics 310, 37−53.

Zhao, G.C., Cawood, P.A., Wilde, S.A., Sun, M., Lu, L.Z., 1999b. Thermal evolution of two textural types of mafic granulites in the North China Craton: evidence for both mantle plume and collisional tectonics. Geol. Mag. 136, 223−240.

Zhao, G.C., Cawood, P.A., Lu, L.Z., 1999c. Petrology and P−T history of the Wutai amphibolites: implications for tectonic evolution of the Wutai Complex, China. Precambrian Res. 93, 181−199.

Zhao, G.C., Cawood, P.A., Wilde, S.A., Sun, M., 2000a. Metamorphism of basement rocks in the central zone of the North China Craton: implication for Palaeoproterozoic tectonic evolution. Precambrian Res. 103, 55−88.

Zhao, G.C., Wilde, S.A., Cawood, P.A., Lu, L.Z., 2000b. Petrology and P−T path of the Fuping mafic granulites: implications for tectonic evolution of the central zone of the North China Craton. J. Metamorphic Geol. 18, 375−391.

Zhao, G.C., Wilde, S.A., Cawood, P.A., Sun, M., 2001a. Archean blocks and their boundaries in the North China Craton: lithological, geochemical, structural and P−T path constraints and tectonic evolution. Precambrian Res. 107, 45−73.

Zhao, G.C., Cawood, P.A., Wilde, S.A., Lu, L.Z., 2001b. High-pressure granulites (retrograded eclogites) from the Hengshan Complex, North China Craton: petrology and tectonic implications. J. Petrol. 42, 1141−1170.

Zhao, G.C., Cawood, P.A., Wilde, S.A., Sun, M., 2002a. Review of global 2.1−1.8 Ga orogens: implications for a pre-Rodinia supercontinent. Earth Sci. Rev. 59, 125−162.

Zhao, G.C., Sun, M., Wilde, S.A., 2002b. Major tectonic units of the North China Craton and their Paleoproterozoic assembly. Sci. China 32, 538−549 (Series D Earth Sciences, in Chinese).

Zhao, G.C., Wilde, S.A., Cawood, P.A., Sun, M., 2002c. SHRIMP U−Pb zircon ages of the Fuping Complex: implications for accretion and assembly of the North China Craton. Am. J. Sci. 302, 191−226.

Zhao, G.C., Sun, M., Wilde, S.A., Li, S.Z., 2003. Assembly, accretion and breakup of the Paleo-Mesoproterozoic Columbia Supercontinent: records in the North China Craton. Gondwana Res. 6, 417−434.

Zhao, G.C., Sun, M., Wilde, S.A., Li, S.Z., 2004a. A Paleo-Mesoproterozoic supercontinent: assembly, growth and breakup. Earth Sci. Rev. 67, 91−123.

Zhao, G.C., Sun, M., Wilde, S.A., Li, S.Z., 2005. Late Archean to Paleoproterozoic evolution of the North China Craton: key issues revisited. Precambrian Res. 136, 177−202.

Zhao, G.C., Sun, M., Wilde, L.i.,S.Z., Liu, S.W., Zhang, J., 2006. Composite nature of the North China Granulite-Facies Belt: tectonothermal and geochronological constraints. Gondwana Res. 9, 337−348.

Zhao, G.C., Kröner, A., Wilde, S.A., Sun, M., Li, S.Z., Li, X.P., et al., 2007. Lithotectonic elements and geological events in the Hengshan-Wutai-Fuping belt: a synthesis and implications for the evolution of the Trans-North China Orogen. Geol. Mag. 144, 753−775.

Zhao, G.C., Wilde, S.A., Sun, M., Guo, J.H., Kröner, A., Li, S.Z., et al., 2008a. SHRIMP U−Pb zircon geochronology of the Huaian Complex: constraints on late Archean to Paleoproterozoic crustal accretion and collision of the Trans-North China Orogen. Am. J. Sci. 308, 270−303.

Zhao, G.C., Wilde, S.A., Sun, M., Li, S.Z., Li, X.P., Zhang, J., 2008b. SHRIMP U−Pb zircon ages of granitoid rocks in the Lüliang Complex: implications for the accretion and evolution of the Trans-North China Orogen. Precambrian Res. 160, 213−226.

Zhao, G.C., He, Y.H., Sun, M., 2009c. Xiong'er volcanic belt in the North China Craton: implications for the outward accretion of the Paleo-Mesoproterozoic Columbia (Nuna) Supercontinent. Gondwana Res. 16, 170−181.

Zhao, G.C., Wilde, S.A., Guo, J.H., Cawood, P.A., Sun, M., Li, X.P., 2010a. Single zircon grains record two Paleoproterozoic collisional events in the North China Craton. Precambrian Res. 177, 266−276.

Zhao, G.C., Wilde, S.A., Zhang, J., 2010c. New evidence from seismic imaging for subduction during assembly of the North China Craton: comment. Geology 38, e206.

Zhao, G.C., Li, S.Z., Sun, M., Wilde, S.A., 2011a. Assembly, accretion, and break-up of the Palaeo-Mesoproterozoic Columbia supercontinent: records in the North China Craton revisited. Int. Geol. Rev. 53, 1331−1356.

Zhao, G.C., Cawood, P.A., Wilde, S.A., Sun, M., Zhang, J., He, Y.H., et al., 2012. Amalgamation of the North China Craton: key issues and discussion. Precambrian Res. 222−223, 55−76.

Zhao, L., Zheng, T.Y., Chen, L., Ai, Y.S., He, Y.M., Lu, G., 2011b. Lithosphere−asthenosphere interactions and geodynamic cause of the North China Craton destruction: constraints from seismic observations. The International Conference on Craton Formation and Destruction (ICCFD), Beijing, Abstract.

Zhao, R.F., Guo, J.H., Peng, P., Liu, F., 2010b. 2.1 Ga crustal remelting event in Hengshan Complex: evidence from zircon U−Pb dating and Hf−Nd isotopic study on potassic granites. Acta Petrol. Sin. 27, 1607−1623.

Zhao, T.P., Zhou, M.F., Zhai, M.G., Xia, B., 2002d. Paleoproterozoic rift-related volcanism of the Xiong'er group, North China Craton: implications for the breakup of Columbia. Int. Geol. Rev. 44, 336−351.

Zhao, T.P., Zhai, M.G., Xia, B., Li, H.M., Zhang, Y.X., Wan, Y.S., 2004b. Zircon U−Pb SHRIMP dating for the volcanic rocks of the Xiong'er Group: constraints on the initial formation age of the cover of the North China Craton. Chin. Sci. Bull. 49, 2495−2502.

Zhao, T.P., Zhou, M., Zhai, M.F., Xia, B., 2004c. Single zircon U−Pb ages and their geological significance of the Damiao anorthosite complex, Hebei Province. Acta Petrol. Sin. 20, 685−690.

Zhao, T.P., Chen, W., Zhou, M.F., 2009b. Geochemical and Nd−Hf isotopic constraints on the origin of the ∼1.74-Ga Damiao anorthosite complex, North China Craton. Lithos 113, 673−690.

Zhao, Z.P., Zhai, M.G., Wang, K.Y., Yan, Y.H., Guo, J.H., Liu, Y.G., 1993. Precambrian Crustal Evolution of Sino-Korean Paraplatform. Science Press, Beijing, pp. 366–384 (in Chinese).

Zhao, Z.R., Song, H.X., Shen, Q.H., Song, B., 2008c. Geological and geochemical characteristics and SHRIMP U–Pb zircon dating of the Yinglingshan granite and its xenoliths in Yishui County, Shandong, China. Bull. Geol. 27, 1551–1558.

Zhao, Z.R., Song, H.X., Shen, Q.H., Song, B., 2009a. Petro-geochemical characters and SHRIMP U–Pb zircon ages of meta-mafic rocks from Yishui Complex in Yishui County, Shandong Province. Geol. Rev. 55, 287–299.

Zhao, Z.R., Song, H.X., Shen, Q.H., Song, B., 2013. Zircon SHRIMP U–Pb dating of ultramafic rock and mafic granulite from Qinglongshan in Yishui County Shandong Province. Acta Petrol. Sin. 29, 551–563.

Zheng, J.P., Griffin, W.L., O'Reilly, S.Y., Lu, F.X., Wang, C.Y., Zhang, M., et al., 2004a. 3.6 Ga lower crust in central China: new evidence on the assembly of the North China Craton. Geology 32, 229–232.

Zheng, J.P., Griffin, W.L., O'Reilly, S.Y., Lu, F.X., Yu, C.M., Zhang, M., et al., 2004b. U–Pb and Hf-isotope analysis of zircons in mafic xenoliths from Fuxian kimberlites: evolution of the lower crust beneath the North China Craton. Contrib. Mineral. Petrol. 148, 79–103.

Zheng, P.X., Jin, W., Zhou, Y., Li, J., Zheng, C.Q., 2009a. Zircon U–Pb age and geological significance of the Taizili Granitic Gneiss from Western Liaoning Province. J. Jilin Univ. 39, 454–460 (Earth Science edition).

Zheng, T.Y., Zhao, L., Zhu, R.X., 2009b. New evidence from seismic imaging for subduction during assembly of the North China Craton. Geology 37, 395–398.

Zheng, T.Y., Zhao, L., Zhu, R.X., 2010. New evidence from seismic imaging for subduction during assembly of the North China Craton: reply. Geology 38, e207.

Zheng, T.Y., Zhu, R.X., Zhao, L., Ai, Y.S., 2012. Intralithospheric mantle structures recorded continental subduction. J. Geophys. Res. Solid Earth 117. Available from: 10.1029/2011JB008873.

Zhong, C.T., 2010. Evolution of the End-Neoarchean to Paleoproterozoic adakites and sanukitoids from the Daqingshan-Yinshan area, Inner Mongolia and its constraints on mineralization. Final report of the "Hundred Talent Scheme" research, Ministration of Lands and Resources, pp. 1–13.

Zhou, J.B., Zheng, Y.F., Yang, X.Y., Su, Y., Wei, C.S., Xie, Z., 2002. Tectonic framework and evolution of the Bayan Obo area. Geol. J. Chin. Univ. 8, 46–61.

Zhou, X.W., Geng, Y.S., 2009. Metamorphic age of the khondalite series in the Helanshan region: constraints on the evolution of the western block in the North China Craton. Acta Petrol. Sin. 25, 1843–1852.

Zhou, X.W., Zhao, G.C., Wei, C.J., Geng, Y.S., Sun, M., 2008. Metamorphic evolution and Th–U–Pb zircon and monazite geochronology of high-pressure pelitic granulites in the Jiaobei massif of the North China Craton. Am. J. Sci. 308, 328–350.

Zhou, X.W., Zhao, G.C., Geng, Y.S., 2010. Helanshan high-pressure pelitic granulites: petrological evidence for collision event in the Western Block of the North China Craton. Acta Petrol. Sin. 26, 2113–2121.

Zhu, R.X., Zheng, T.Y., 2009. Destruction geodynamics of the North China Craton and its Paleoproterozoic plate tectonics. Chin. Sci. Bull. 19, 3354–3366.

Ziegler, P.A., Cloetingh, S., van Wees, J.-D., 1995. Dynamics of intra-plate compressional deformation: the Alpine foreland and other examples. Tectonophysics 252, 7–59.

Printed and bound by CPI Group (UK) Ltd, Croydon, CR0 4YY

03/10/2024

01040421-0018